ADVANCED CERAMICS III

ADVANCED CERAMICS SERIES

Editor: Shigeyuki Sōmiya

Contents of Previous Volume
Advanced Ceramics II

ADVANCED CERAMICS III

Edited by

SHIGEYUKI SŌMIYA

Dean Professor, The Nishi Tokyo University, Japan
Professor Emeritus, Tokyo Institute of Technology, Japan

ELSEVIER APPLIED SCIENCE
LONDON and NEW YORK

ELSEVIER SCIENCE PUBLISHERS LTD
Crown House, Linton Road, Barking, Essex IG11 8JU, England

Sole Distributor in the USA and Canada
ELSEVIER SCIENCE PUBLISHING CO., INC.
655 Avenue of the Americas, New York, NY 10010, USA

WITH 25 TABLES AND 137 ILLUSTRATIONS

© 1990 ELSEVIER SCIENCE PUBLISHERS LTD
Softcover reprint of the hardcover 1st edition 1990

British Library Cataloguing in Publication Data

Advanced ceramics III.
1.   Ceramics
I.   Sōmiya, Shigeyuki
666

ISBN-13: 978-94-010-6829-1

Library of Congress Cataloging-in-Publication Data

Advanced ceramics III / edited by Shigeyuki Sōmiya.
    p.   cm.—(Advanced ceramics series)
    Includes bibliographical references.
    ISBN-13: 978-94-010-6829-1   e-ISBN-13: 978-94-009-0763-8
           DOI: 10.1007/978-94-009-0763-8
    1. Ceramics.   2. Ceramic materials.   I. Sōmiya, Shigeyuki.
II. Title: Advanced ceramics 3.   III. Series.
TP807.A327   1990
666—dc20

No responsibility is assumed by the publisher for any injury and/or damage to persons or property as a matter of products liability, negligence or otherwise, or from any use or operation of any methods, products, instructions or ideas contained in the material herein.

Special regulations for readers in the USA

This publication has been registered with the Copyright Clearance Center Inc. (CCC), Salem, Massachusetts. Information can be obtained from the CCC about conditions under which photocopies of parts of this publication may be made in the USA. All other copyright questions, including photocopying outside the USA, should be referred to the publisher.

To
My Professors
My International Colleagues
and
My International Friends

The late Professor Emeritus Ikutaro Sawai
Professor Emeritus Toshiyoshi Yamauchi
Professor Emeritus Yunoshin Imai
Professor Emeritus Tokichi Noda
Professor Emeritus E. F. Osborn
Professor Emeritus J. A. Pask
Professor Emeritus Megumi Tashiro
Professor Emeritus Shinroku Saito
Professor Arnulf Muan
Professor Rustum Roy
Professor G. Petzow
Professor F. P. Glasser
Dr J. B. MacChesney
Professor R. M. Spriggs
Professor R. F. Davis
Professor Nils Claussen

Without their advice and cooperation this record of international activities would not have been possible.

Foreword

This volume is one in a series which attempts to bring together comprehensive articles on recent advances in ceramics.

The volume is dedicated to Professor Shigeyuki Sōmiya on the occasion of his retirement from the Tokyo Institute of Technology; and it is a most fitting tribute. Professor Sōmiya has been one of the earliest and most persistent and versatile champions of research in ceramic materials in Japan. He has served this cause extraordinarily well by mixing two strategies. First, by making bridges to the entire international community of ceramic researchers in the US and Europe. Thereby, he kept a window for all of Japanese ceramic science on world class research in the field. It was largely through his efforts that the series of US—Japan International Cooperation Sessions in Ceramics were started. I was honored to be US chairman of the first such in 1969. At Penn State we are delighted to claim Professor Sōmiya as an honorary alumnus. The high regard in which he is held is shown by the many of his colleagues from the University who have chosen to come over for this conference. He was also recognized with a Penn State MRL Bridge-Building award in 1988 to reflect his pioneering in establishing the two-way exchange with Japan.

Second, by his own research, many of the papers which are presented in this volume correspond to Sōmiya's interest and emphasis. I first note that they are all concerned with synthesis and preparation of materials. This forms Sōmiya's life work—making new materials using new processes such as hydrothermal reactions. I also note that when much of this was being done the field of synthesis was totally out of fashion. Color centers in alkali halides, diffusion in MgO, work to fracture and a wide variety of physical measurements on single crystals were all the rage. Yet in 1989 the US National Academy of Engineering and of Sciences, in their report on Materials Science and Engineering, clearly identify 'synthesis' as the major neglected area of opportunity.

This volume forms a very worthwhile record of the importance of the field of new materials syntheses and processes ranging from fine powders to hydrothermal to cements to electroceramics to diamonds.

RUSTUM ROY

Preface

This book is one of the series of Advanced Ceramics of which I am editor. The first one, *Advanced Ceramics I*, was published in 1987 by the Terra Scientific Publishing Company, Tokyo. The second, *Advanced Ceramics II*, was published by Elsevier Applied Science Publishers, UK in 1988. This is the third book in the series.

The meeting for *Advanced Ceramics III* was held on May 27, 1988 in Tokyo as a commemorative meeting, marking my retirement from the Tokyo Institute of Technology at 60 years of age.

Speakers, session chairmen and chairwomen at this meeting were distinguished professors and engineers working at the frontiers of ceramic science and technology, and who have helped the science and technology of ceramics to grow. Also they are good friends and colleagues of mine. They were, by no means, all the familiar faces from the world of ceramics, but the speakers covered the most interesting fields in ceramics at that present time.

Synthesis of materials is one of the key issues in ceramics today. This book is one of several comprehensive guides to ceramic science and technology, especially concerning the synthesis of materials. In particular, several papers are related to hydrothermal research and synthesis.

I studied the synthesis of minerals related to chrom-magnesia refractories, in the 1920s under the guidance of Professor Toshiyoshi Yamauchi from the Tokyo Institute of Technology. Phase diagram studies were carried out under the guidance of Professor E. F. Osborn and Professor Arnulf Muan at the Pennsylvania State University. They

ix

have shown how important materials synthesis is! This book will be one of the books which solves the problem of materials synthesis.

Academic advice during my studies was provided by Professor Toshiyoshi Yamauchi, Professor Shinroku Saito, Professor Rustum Roy and Professor J. A. Pask. Academic advice related to experimental techniques was provided by Professor Hiroshige Suzuki and Professor Arnulf Muan.

During my study at the Pennsylvania State University I recognised the importance of international cooperation with contributions from Professor E. F. Osborn, Professor Rustum Roy, Professor Arnulf Muan, Professor Della M. Roy. After this study abroad good advice was given by Professor J. A. Pask and Professor R. M. Fulrath of University of California, Berkeley CA, Professor Rustum Roy of the US and Professor G. Petzow, Germany. Studying abroad at the Pennsylvania State University from 1957 to 1959 and at the Max-Planck Institut für Metallforschung, Pulvermetallgisches Laboratorium in 1977 was very beneficial and I very much enjoyed life at these institutions.

More than 600 persons attended this meeting. As well as the participants from Japan, people came to Japan from the US, Europe, Korea, China, and around the world. I have learnt many things from them.

After reading this book, you will be able to understand the art, science and technology of ceramics. If you would like to know more about this subject, I recommend the following books published by Elsevier Applied Science Publishers, UK: *Sintering '87* (1988), *Sintering, Key Papers* (1990), *Hydrothermal Reactions for Materials, Science and Engineering, An Overview of Research in Japan* (1990), *Silicon Nitride—1* (1990), *Silicon Carbide, Vols 1 and 2* (1991). Also, *Advances in Ceramics, Vol. 24, Zirconia III* (1988) published by the American Ceramic Society, *Fundamental Structural Ceramics* (1987) published by the Terra Scientific Publishing Company, Tokyo and *Advanced Technical Ceramics* (1989) published by Academic Press, USA. These books are all written in English but show the Japanese standard of ceramics study.

I am grateful to the speakers, chairmen and chairwomen of the lectures and the participants of the meeting. Finally many foundations, companies and participants contributed to this meeting. Without these contributions and my study abroad at the Pennsylvania State University, University Park, PA, USA and Max-Planck Institute für Metallforschung, Pulvermetallgisches Laboratorium, Stuttgart, FRG, I would not have been able to organize the meeting and generate interest for international cooperation, exchanging data and ideas formally and informally with others from around the world. Also, without the benefit of study abroad, I would not have been able to commence my studies on phase equilibria and hydrothermal reactions.

SHIGEYUKI SŌMIYA

Acknowledgements

1. I am grateful to the speakers, chairmen and chairwomen, and more than 600 participants of the meeting.

2. I wish to express my appreciation to the following people, foundations and companies:

Arumu Publishers
Asahi Glass Col. Ltd
Asahi Glass Kogyo Gijutsu
 Shoreikai

Bilingual Group

Central Glass Co. Ltd
Concrete Pole Pile Association
Colloide Research Co. Ltd
Chichibu Cement Co. Ltd
Chunichi Newspapers Co. Ltd

Dainippon Ink Chemical Co. Ltd
Daiichi Kigenso Chemical Industries Co. Ltd

Daiichi Cement Co. Ltd.
Denki Kagaku Kogyo K.K.
Dowa Mining Co. Ltd

Elsevier Science Publishers Ltd
Eiko Seiki Co. Ltd
Eru Liquid Laboratories Co. Ltd

Fukuya Shoten
Fujikura Densen K.K.
Fine Ceramic Fair Kaisai
 Kyogikai
France–Japan Kogyo Gijutsu
 Kyokai

Gihodo Shuppan Co. Ltd

Hitachi Ltd
Honda Gijutsu Kenkyusho K.K.
Hayakawa Rubber Co. Ltd
Harcourt Brace Jovanovich Japan Inc.
Hitachi Kasei Kogyo Co. Ltd
Harima Refractories Co. Ltd

Inoue Japax Co. Ltd
ICI Japan Co. Ltd

Japan Fine Ceramic Association
JEOL Ltd

Kobe Steel Corp.
Kyoritsu Publishing Co. Ltd
Kojundo Chemicals Co. Ltd
Koyosha Co. Ltd
Kawasaki Refractories Co. Ltd
KSP Co. Ltd
KDP Co. Ltd
Kyocera Con.
Kureha Chemical Industries Co. Ltd
Kurosaki Refractories Co. Ltd
Kawasaki Steel Corp.
Kyokai Tsushinsha Co. Ltd
Koransha Co. Ltd

Mitsui Toatsu Kagaku Co. Ltd
Mitsubishi Metals Co. Ltd
Maeda Seikan Co. Ltd
Murata Seisakusho Co. Ltd
Mitsui Metals and Mining Co. Ltd
Mitsui Trading Co. Ltd, Denshi Hambai Div.

Morimura Brothers Inc.
Mino Refractories Co. Ltd

Nihon Cement Co. Ltd
Nikkiso Co.
Nisshin Kagaku Co. Ltd
Nitto Denki Industries Co. Ltd
Nippon Ui Tech Co. Ltd
Nippon Karito Co. Ltd
Nippon Ferro Co. Ltd
Nippon Res & Dev. Co. Ltd
NKK Co.
NEC Co. Ltd
NGK Spark Plug Co.
Nippon Itagarasu Zairyo Kogaku Joseikai
Nikkan Kogyo Shimbun Co. Ltd
Nissan Motors Co. Ltd
Nippon ITS Co. Ltd
NGK Insulators Ltd
Nippon Ceramics Co. Ltd
Nippon Chemical Ceramics Co. Ltd
Nippon Mining Co. Ltd
Nippon Soda Co. Ltd
Nippon Renga Seizo Co. Ltd
Nisshin Kokuen K.K.

Overseas X-ray Service Co. Ltd
Optron Co. Ltd
Okura Riken Co. Ltd
Onoda Cement Co. Ltd
Osaka Package Co. Ltd
Okura Denki K.K.

Riko Shinsha K.K.
Rhone-Poulenc Japan Co. Ltd
Rigaku Denki K.K.

Sankei Co. Ltd
Sakai Kagaku Kogyo K.K.
Sumitomo 3M Co. Ltd
Sumitomo Cement Co. Ltd
Sumitomo Metal and Mining Co. Ltd
Sumitomo Electrical Industries Co. Ltd.
Showa Denko K.K.
H. C. Starck Far East Co. Ltd
Seimi Chemical Co. Ltd
Sumitomo Trading Co. Ltd
Shin-etsu Chemical Ins.
Senyo Glass Industries Co.
Suzuki Motors Co. Ltd
Shinagawa Refractories Co. Ltd
Suzuki Riken Kogyo Co. Ltd

Tokyo Koki Co. Ltd
Taiyo Yuden Co. Ltd
TEP Co. Ltd
Tori Chemical Laboratory Co. Ltd
Token Sangyo Co. Ltd
Teijin Co.
Tosoh Co. Ltd
Tokai Konetsu Industries Co. Ltd
Toshiba Tungaloy Co. Ltd
Tokuyama Soda Co. Ltd
Takeda Printing Co. Ltd
Toray Co. Ltd
Tokyo Kogyo Co. Ltd
Toto Ltd

Ushio Denki Co. Ltd
Ushio U-Tech Co. Ltd
Uchida Publishing Co. Ltd

Yamaka Seito Co. Ltd

There were about 600 people who participated in the meeting and more than a thousand who made contributions. Unfortunately I am unable to mention all these individuals by name but I am, however, very grateful to all those concerned.

Contents

List of Contributors

L. E. CROSS
Evan Pugh Professor of Electrical Engineering, Director of the Materials Research Laboratory, The Pennsylvania State University, University Park, Pennsylvania 16802, USA.

R. C. DEVRIES
17, Van Vorst Drive, Burnt Hills, New York 12027, USA.

F. P. GLASSER
Department of Chemistry, University of Aberdeen, Meston Walk, Old Aberdeen, AB9 2UE, Scotland.

F. F. LANGE
Materials Department, College of Engineering, University of California, Santa Barbara, California 93106, USA.

J. B. MACCHESNEY
AT & T Bell Laboratories, 600 Mountain Avenue, Murray Hill, New Jersey 07974, USA

A. MUAN
Department of Geosciences, The Pennsylvania State University, University Park, Pennsylvania 16802, USA.

G. PETZOW
Max-Planck-Institut für Metallforschung, Institut für Werkstoffwissenschaft, Pulvermetallurgisches Laboratorium, Heisenbergstr. 5, D-7000 Stuttgart 80, FRG.

THE LATE A. RABENAU
Max-Planck-Institut für Festkörperforschung, D-7000 Stuttgart 80, FRG.

R. ROY
Materials Research Laboratory, The Pennsylvania State University, University Park, Pennsylvania 16802, USA.

H. SCHUBERT
Max-Planck-Institut für Metallforschung, Institut für Werkstoff-wissenschaft, Pulvermetallurgisches Laboratorium, Heisenbergstr. 5, D-7000 Stuttgart 80, FRG.

S. SŌMIYA
The Nishi Tokyo University, 3-7-19 Seijo, Setagaya, Tokyo 157, Japan.

B. C. H. STEELE
Centre for Technical Ceramics, Imperial College, Cromwell Road, London, SW7 2BA, UK.

1

A Strategy for Research on Synthesis of Ceramics Materials

Rustum Roy

Materials Research Laboratory, The Pennsylvania State University, University Park, Pennsylvania, USA

ABSTRACT

Advances in materials research are closely correlated with the discovery or purposive synthesis of new materials or new processes for preparing materials. During the last two or three decades, there has been emerging a capacity to 'design' materials optimized for a particular use or device, and to synthesize new materials to these specifications. However, serendipity still plays a major role in new materials development, and the question of the optimum research strategy for materials synthesis studies has received insufficient attention. Such research strategy may be just as important as new instruments in the search for new materials.

A brief review of novel syntheses and new processes across the whole field of ceramics is followed by examples from the author's laboratory of the synthesis of new materials involving unit cell level manipulation of composition, and changes at the macromolecular level and at the nanometer level. The new materials discussed range from zero-expansion ceramics to nanocomposites and superconductors and new processes range from those for making powders to those for ultra-high pressure phases. The case studies illustrate the author's research strategy, combining state-of-the-art empirical 'theory' with opportunistic response to serendipitous observations.

1. INTRODUCTION

This paper attempts to address the question whether, and if so how, we can consciously set out to create genuinely new materials, and

1

especially such materials with pre-selected (desirable) properties. Is materials synthesis only an art ruled largely by chance, or is it now reduced to a logical algorithm? We start by examining the empirical state of affairs.

In recent years the volume of research conducted in a particular field of materials is typically not a function of the scientific opportunity, the innovativeness of the concepts, nor technological potential of the subject. Instead, research efforts especially in academia, but partly in industry, appear to follow other forces:

(1) Fashions and trends set in the public media and the professional press;
(2) momentum and inertia accreted around a group (especially if based on (3));
(3) specialized large pieces of equipment that demand to be fed not only money but problems;
(4) funding agency emphases or requests for proposals (RFPs).

Illustrative evidence for this may be clearly seen in the data in Table 1 on three very recent, major discoveries in new materials synthesis. In this table (taken from Ref. 1) I compare the relevant parameters for the new high-T_c superconductors, the metastable CVD diamond films, and the new ceramic-in-ceramic composite, Lanxide, all announced to the same society meeting audiences. One notes that Lanxide, with radically new science and technology (1200 patents filed), has attracted virtually no materials researchers, although hundreds specialize in ceramic composites. Conversely, oxide superconductor research based on an unexpected scientific advance in an old and mature field of technology that had had little growth for two decades has attracted literally thousands of amateurs and non-specialists in materials synthesis to dabble in the field. CVD diamonds, also based on an equally unexpected scientific advance, even after its demonstrated technological potential in one country (Japan) fails to attract large numbers of active researchers in others. Why? The explanation lies clearly in the row of Table 1 describing the extent of national publicity attending each of these discoveries. The *perception,* created by the public press, partly by individual scientists, and echoed in the science press, is that there is some enormous *technological* payoff even in the 10-year time frame. Even in the professional society press it is not made clear that such claims have been contradicted by every consensual examination of the issue by experts [2]. What we learn from this case study is that

TABLE 1
Comparison of three recent major new discoveries in new materials synthesis

	Three discoveries (starting) at		
	MRS Dec. 85 Lanxide announced	MRS Dec. 86 High-T_c super- conductor confirmed	MRS Dec. 87 Diamond films announced and confirmed
Advertising hype	Strictly avoided Technical press missed it	Enormous	Very minor Technical press picked it up
Impact on research	Not one proposal submitted to NSF, DOE	Cast of thousands, working outside their field	Considerable interest. Minor initiatives
Impact on agencies	Only DoD response Ignored by ALL others	Measured, studied response. *Not* overwhelmed	DoD response; others ignore
Impact on industrial research	Major potential Replacement and new technologies	Unknown Most analyses— dubious	Major potential on wide variety of existing products
U.S. interest	100% U.S.	U.S. role—minor	U.S. role—very small
Technological position	>1200 patents filed by one U.S. company	Patent positions— appear irrelevant so far	Japanese far ahead in technology, USSR in science

scientists are like gamblers and lottery players, driven principally by the promise of extraordinarily high payoffs, even if these are largely illusory.

The fact that many major advances in the discovery of new materials occur as a result of discriminating and perceptive observation of serendipitous events leads one to a relatively clear choice of strategies in conducting research in this field. This strategy can be summarized as follows. When applied to innovative synthesis of new single phases, i.e. to the design of solids at the unit cell level:

(1) Substitute 'chemistry + structure' as twin determinants of properties instead of the hackneyed yet obviously incorrect 'structure–property' connection.

(2) Be thoroughly grounded in state-of-the-art *crystal chemistry* and *phase equilibrium* as the essential theoretical base to design systematic experiments.

(3) Use a wide range of synthetic tools covering the range from liquid He to a few thousand Kelvins in temperatures, and from vacuum to a hundred kilobars or so in pressure. (Most materials research laboratories with millions of dollars in analytical tools have hardly any materials-preparation tools other than mortars, ball mills and 1-atmosphere pressure furnaces!)

(4) Thus equipped, while one establishes a systematic search—let us say for a new ferrimagnetic material with three sublattices—one must be prepared to follow totally unexpected leads or hints picked up in the course of the systematic work, or from the worldwide journal and patent literature, or from visits to other laboratories anywhere in the world.

The subsequent material is presented as a series of 'case studies' of particular issues.

1.1. New Materials: A Comparison of Ferroelectrics. Ferrimagnets and High-T_c Superconductors

Table 2 lists a comparison of the facts surrounding three major new families of ceramic materials that were discovered in the period 1945–1988. This set is an excellent group for learning from retrospective analysis what one might do in the future.

The first difficult question that arises is: On what scale does one measure the scientific significance of an advance in new materials properties? Which was a bigger step function—going from a $K = 10$ for mica to a $K \approx 10\,000$ for $BaTiO_3$ or going from a T_c of 23 K for Nb_3Ge to 36 K for $LaBaCuO_4$ (or 90 K for $YBa_2Cu_3O_{6.5}$)?

(1) One criterion of course is: What impact did it have on the neighboring sciences? (Weinberg, Ref. 3).

(2) A second is: What impact is it likely to have on dependent or future technologies?

Science. The discovery of $BaTiO_3$ (Wainer *et al.*; Wul and Goldman; see Ref. 4) triggered an enormous field of the science of ferroelectrics, and eventually all ferroics, transducers, electro-optic materials, etc.

The discovery of rare-earth garnets by Keith and Roy in 1954 [5], on the other hand, remained confined to ferrimagnetism, in which theories developed in spinel ferrites were expanded and elaborated to the three sublattice garnets. While very little developed in neighboring

TABLE 2

Comparison of history of some major advances in properties caused by discovery of new materials

	Three discoveries (past) Comparative research history & strategy		
	'Ferroelectricity' TiO_2 $MgTiO_3$ $BaTiO_3$	'Superconductivity' $LaBaCuO$ $YBa_2Cu_3O_7$	'Ferrimagnetism' $Sm_3Ga_2Ga_3O_{12}$ $Y_3Fe_2Fe_3O_{12}$
Discovery	USA ~Simultaneously: USA, USSR, Japan 1945–46	Switzerland–Japan ~Simultaneously: Huntsville, Peking, Bangalore 1986–87	USA ~Simultaneously: BTL, Grenoble 1954–56
Property advance	$K = 10$ $\to 100$ $\to 10,000$	$T_c = 23$ K $\to 36$ $\to 90$	Max $\mu_B = 10$ $\to 36$
Research effort[a]	100 p.y. $\to$ major payoff	(1000 p.y. and far to go to??)	100 p.y. $\to$ minor payoff
U.K. position at time of discovery	Industry in commanding position	Industry very weak Universities very weak in relevant fields	Industry far ahead. Still small payoff
Technological prospects	*Drop-in replacement* into existing products	~No products Short-term prospects unclear	No impact on ferrites. Memory applications trivial. Microwave—minor.
Technological reality	Enormous capacitor industry Piezoelectrics, pyroelectrics, electro-optics	Major impact Improbable in 10–15-year frame	Impact—minor

[a] p.y. = person years.

fields, the excellent work of the Néel school greatly deepened our knowledge of ferrimagnetism.

High-T_c superconductivity has likewise triggered enormous activity in superconductivity theory but whether or not any technology even comes out of it, it is unlikely to spinoff beyond superconductivity.

Technology. $BaTiO_3$ and its closely related perovskite materials in a decade transformed an existing industry—the capacitor industry. From that it led to the PZT-based transducer industry and later to the electro-optic materials. There was no doubt, immediately after its discovery, that it would transform a whole industry. New transducers were not foreseen, and some *potential* applications, such as information storage ran decades ahead of their realization.

The rare-earth garnets had very little application in the magnetics industry, although eventually $Y_3Fe_2Fe_3O_{12}$ (YIG) carved out a niche for itself in microwave antennas. The promise of bubble memory storage, which was actively studied for years, was never realized.

The perovskite and K_2NiF_4 superconductors are new materials with a modest advance in one property (T_c) with substantial deficiencies (so far) in others. Superconductivity, unlike ferroelectricity, had been thoroughly studied and utilized in several devices and commercial applications for 2–3 decades. This has led to some rather minor industries (e.g., magnetic resonance imaging) and prototyping of others (generators, etc.) but no great opportunities have appeared. The new materials cannot yet even be substituted in these applications, let alone advance them.

A major advance in materials synthesis can therefore be described as:

- one which brings a new phenomenon to bear in meeting the needs of a traditional function and/or
- one which exhibits a step-function advance (by an order of magnitude?) in the critical property so that it can be substituted with major advantage *in an existing technology*.

By these standards one can see that $BaTiO_3$ was a truly revolutionary new material when it was discovered, and perhaps did not receive the research attention it deserved. Conversely, the high-T_c superconductors were a modest discovery upon which a great excess of attention has already been bestowed.

1.2. New Processes: Comparison of Solution–Sol–Gel (SSG),† Glass-Ceramic and Transformation Toughening Processes

Science. The scientific basis for mixing in solution, forming sols and thence gels, and the use of organic precursors and inorganic sols to make homogeneous ceramics was fully developed in some dozen papers and hundreds of compositions by the author and his students in the decade following 1948 (see Ref. 6). The ceramic *science* community ignored the discovery for nearly 30 years.

† Please see p. 243 for SSG definitions.

The discovery of the glass-ceramic process was based on Corning's long experience with post-formation heat treatment of glasses. It triggered a modest scientific research effort, technologically driven, to understand the process. This led in turn to the discovery of metastable unmixing and thence to the enormous literature on spinodals and nucleation and crystallization in glasses.

Transformation toughening likewise emerged slowly out of the serendipitously discovered toughness of $Al_2O_3-ZrO_2$ commercial abrasive grain followed by Garvie's observation on fine-grained ZrO_2 [7]. In both the latter two cases, technology definitely preceded science, whereas in the first the science came first.

Technology. SSG was utilized some 15–20 years after its scientific discovery in a series of technologies: making nuclear fuel pellets; making ceramic fibers; making thin-film coatings for windows; and making abrasive grain. These are modest-size technologies which are well known and are constantly looking for new applications. They in turn belatedly gave rise to a major science push to understand the SSG process in great detail, although the chances for major new technologies is very slight.

Glass-ceramics by contrast made an immediate and revolutionary impact on industry in the home cooking appliance field and eventually became a major new ceramic-forming technology world-wide.

Transformation toughening, as its technological potential became widely advertised, generated a great deal of science, and has now generated some small-size technologies.

Since processing necessarily requires scale-up and technological utilization to prove its value, it is difficult to generalize on the optimum approach path. The process innovator must be coupled more closely to the demand or need in industry for effective adaptation to be guided.

2. EXAMPLES OF OUR OWN APPLICATION OF THIS STRATEGY

In the following three or four sections I describe work done in our laboratories with my colleagues and my students on a series of topics, each involving a new material or new process. Our data clearly show that *experimental* experience is the key, together with the ability to recognize genuinely new opportunities which appear serendipitously.

2.1. SSG Route to New Family of Nanocomposites

Starting in 1948 the author developed the science of solution mixing and the sol–gel process to make ultra-*homogeneous* glasses and ceramics. In a series of papers, an enormous number of common ceramic systems were studied and literally hundreds of simple and complex compositions were made by the SSG process, starting with both organic and inorganic precursors (see Refs. 6, 8, 9). Subsequent to the development of several technologies based on the SSG process (see above), a second round of detailed scientific studies was started in the late 1970s. From a strategic point of view, in spite of the fashion which had developed and the availability of funds, I chose to stay out of the field since I knew that the technological potential had been rather thoroughly explored by many research groups world-wide, and no new scientific concepts had emerged. In the early 1980s, however, I found an entirely new direction for SSG research—indeed, it was the opposite of the original direction. Originally, I had decided that mixing in solution was the only way to get true *homogeneity*. The gel route provided the means of preserving that homogeneity through the drying stage. The new concept was to use the SG technique to get 'maximum *heterogeneity*' into a solid. How does one maximize heterogeneity? By increasing the surface area between dissimilar phases, i.e. by mixing the finest possible powders in a truly statistical manner. This cannot be achieved by mixing solid powders but is relatively easily done by mixing sols. Thus was born our new field of di-phasic xerogel research which has proved so novel and fruitful. The materials they form are true *nanocomposites,* i.e. two (or more) phases mixed on a nanometer scale.

Gels may be multi- or di-phasic with respect to *composition* or *structure*. Thus, a mullite gel of $3Al_2O_3 \cdot 2SiO_2$ composition can also be made up as a di-phasic xerogel by mixing two sols, one of Al_2O_3 and one SiO_2, respectively, so that in the xerogel stage there are two separate *phases* each about 20 nm in diameter, one with the composition 'Al_2O_3' and the other 'SiO_2'. The sintering reactions of a compositionally diphasic xerogel are radically different from the sintering reaction of a homogeneous monophasic xerogel, since in addition to the usual minimization of the surface free energy as the driving force, we have the much larger enthalpy of reaction of the two phases.

A *structurally* diphasic mullite gel may be made by adding to a homogeneous mullite composition sol, where the solid is non-crystalline, a second sol with crystalline mullite as the dispersed phase. Although we started the concept of crystallographic seeding of gels nearly 40 years ago [10], the work then was all done in the presence of a fluid phase. The post-1982 work has been all in the solid state.

In a series of papers on phenomena encountered in multi-phasic gels we have been able to demonstrate the following rather remarkable effects.

1. *Greater densification by using compositionally multiphasic gels.* For the system Al_2O_3–SiO_2, Table 3 shows the major difference in sintering caused by using a di-phasic maximally *heterogeneous* instead of a homogeneous single-phase gel. More recently in the system MgO–Al_2O_3–SiO_2, Kijowski, Komarneni and Roy have shown that tri-phasic cordierite gels can be sintered to theoretical density at 1300°C, at which temperature monophasic gels are only 80% dense (see Fig. 1 and Ref. 11).

2. *Lowering reaction temperatures by using structurally diphasic one-component gels.* Figure 2 shows the influence of ~1% of a crystalline gel phase in non-crystalline gels on the transformation temperature of $\theta \rightarrow \alpha$-$Al_2O_3$ (see Ref. 12) and anatase $\rightarrow$ rutile in TiO_2 (see Ref. 13).

3. *Lowering the formation and sintering temperatures by separate and combined compositional and structural di-phasicity.* Table 4 summarizes the major effects one can achieve in a binary compound such as $ZrSiO_4$.

4. *Controlling the desired phases to be formed by providing the corresponding crystallographic seeds in $ThSiO_4$ which has two poly-morphic structures.* Figure 3 shows the extraordinary result that by adding to non-crystalline $ThSiO_4$ sols 1% of a sol of huttonite-

TABLE 3

Sintering behavior of *single*-phase and *diphasic* mullite xerogels as measured by densities

Starting materials	Density at 1200°C	% Theoretical density
Single phase, $Al(NO_3)_3 \cdot 9H_2O$ + $Si(OC_2H_5)_4$	2·71	85·4
Diphasic, $AlOOH$ + $Si(OC_2H_5)_4$	2·92	92·0
Diphasic, $AlOOH$ + Ludox (SiO_2)	3·05	96·2

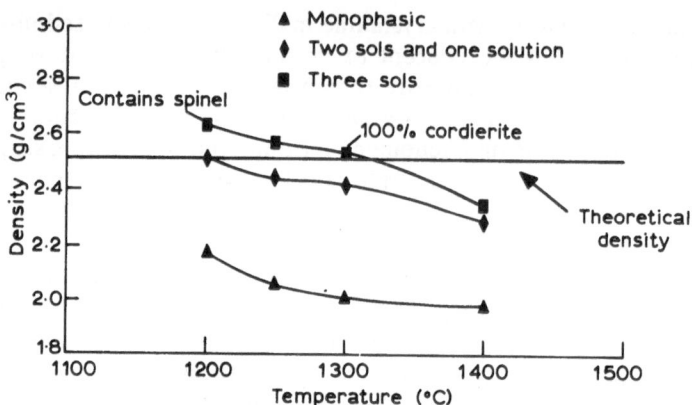

FIG. 1. The difference in sintering of mono-, di- and tri-phasic gels of the cordierite composition. The tri-phasic gel attains theoretical density at 1300°C. (Direct firing at 1400°C produces what may be metastable melting and growth of some large 15 μm crystals.)

structure seeds one can obtain 100% huttonite at 1450°C, whereas side by side an identical sample with 1% thorite seeds yields 100% thorite.

5. *Radical refinement of microstructure by using a structurally di-phasic xerogel.* Figure 4 shows the dramatic change in microstructure in a translucent Al_2O_3 ceramic produced from a single-phase gel and a structurally di-phasic gel respectively; particle size is reduced from 15 μm to <1 μm.

6. *Control of morphology by using doubly di-phasic xerogels.* Figure 5 shows the change in mullite morphology brought about by adding 1% of crystalline mullite seeds to di-phasic mullite xerogel.

2.2. Design of New 'Zero-expansion' Ceramics
In 1947 my colleague F. A. Hummel discovered the family of lithium aluminum silicates—eucryptite and spodumene—which had near-zero thermal expansion [14]. The phase equilibria and crystallization in the system were the subject of my Ph.D. thesis [15]. For nearly the next 40 years this one family dominated the science and technology of low-expansion ceramics, being used in many applications from household dinnerware to 6-m mirror blanks for the largest telescopes.

In the early 1980s we were approached by U.S. Air Force Office of Scientific Research to try to develop a new 'zero-expansion' material with even lower α. While we were not at all optimistic, we accepted

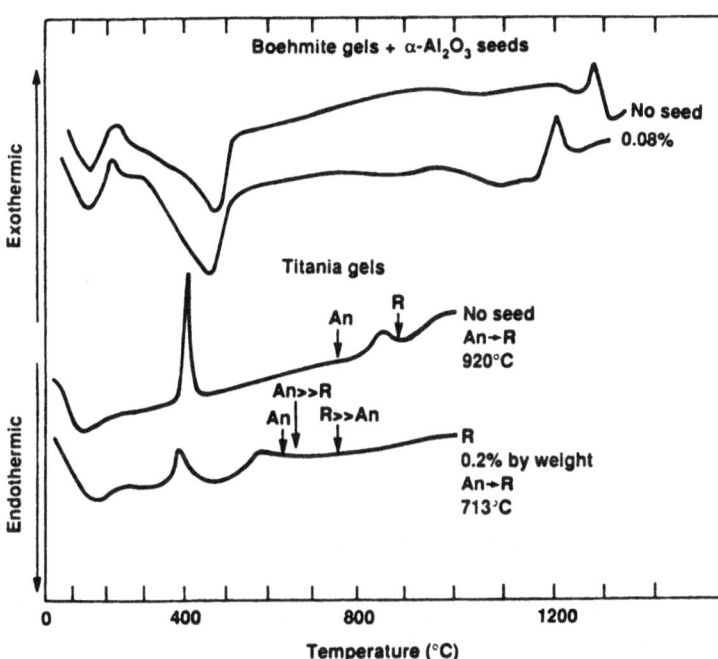

FIG. 2. First recorded instances of solid-state epitaxy. DTA patterns (two upper traces) of non-crystalline single-phase, and structurally diphasic, alumina gels showing the striking lowering of the $\theta \to \alpha$ transformation temperature by the addition of ~1% of 0·1 μm α-Al$_2$O$_3$ crystals. The analogous effect of TiO$_2$-[rutile] seeds on lowering the temperature of the anatase–rutile (An–R) transition in TiO$_2$-[NCS] gels is also shown (two lower traces) (arrows indicate the phases found by X-ray diffraction).

TABLE 4
Crystallization of zircon (ZrSiO$_4$)

	'Homogeneous'[a]	Compositionally diphasic[b]
Without seeds	Formation of ZrSiO$_4$ begins at ≈1350°C	Formation of ZrSiO$_4$ begins at ≈1175°C
Structurally diphasic (with seeds)	Formation of ZrSiO$_4$ begins at ≈1100°C	Formation of ZrSiO$_4$ begins at ≈1075°C

All xerogels treated the same way: 1 h at a given temperature.
[a] Prepared from TEOS and zirconium oxychloride.
[b] Prepared from a silica sol (Ludox) and a tetragonal zirconia sol.

Rustum Roy

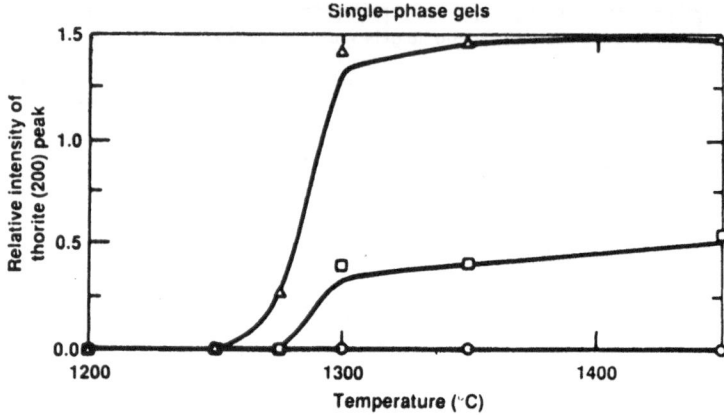

FIG. 3. Percentage of thorite phase formed, by X-ray diffraction, as a function of firing temperature in $ThSiO_4$–[NCS] gels, seeded (made di-phasic) by adding 1% of a sol of $ThSiO_4$–[thorite] crystals ($\triangle$) and by adding 1% of $ThSiO_4$–[huttonite] crystals ($\bigcirc$); unseeded sample ($\square$).

the goal and it became an excellent test case of *how* to synthesize a new material to certain specifications.

Let us first demonstrate how useless the 'composition + structure' vs property relationships are in helping really innovative synthesis. The relationship of thermal expansivity (α) to bond strength is well known [16]. The higher the bond strength the lower the α. But to get zero or negative α would require infinite values of bond strength. Clearly what we were interested in were the anomalies or exceptions to the composition/structure–property relationship. Hence we turned to the always-reliable guide of empirical knowledge. Table 5 shows the list of materials known to have 'low' α. From the commonalities of these materials, I had extracted in the 1950s the guideline for anomalously low expansivities: low-α solids must have strongly bonded polyhedra linked in chain or network open structures which have large 'holes'.

To proceed systematically we were faced with the choices listed below.

ALTERNATIVES FOR DESIGN OF NEW MATERIALS
  A. *Modify single phase*
    • Enormous efforts in Hi-$SiO_{2[qz]}$, $LiAlSiO_4$ not useful
    • Cordierite more promising, especially Ge^{4+} for Si^{4+}
  B. *Design new single phase.* Crystal chemical guidelines available

(a)   1200 deg. C

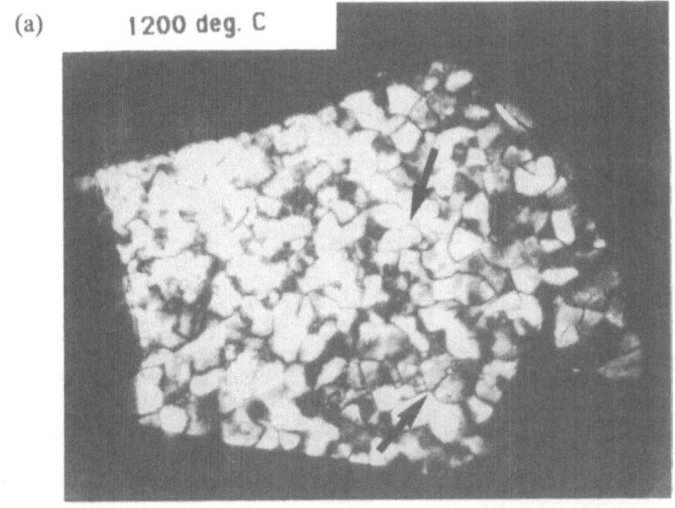

(b)   Seeded with α-Al$_2$O$_3$

FIG. 4. Use of 1% of a second structural phase can *completely* (100%) control the ceramic phase formed even at 1450°C. Figure shows the photomicrograph (crossed Nicol prism) of microstructures of α-Al$_2$O$_3$ formed from boehmite gels heated to 1200°C; (a) single-phase gel; (b) structurally di-phasic gel 'seeded' with α-Al$_2$O$_3$. Note the dramatic grain size refinement from >15 to <1 μm.

(a)

(b)

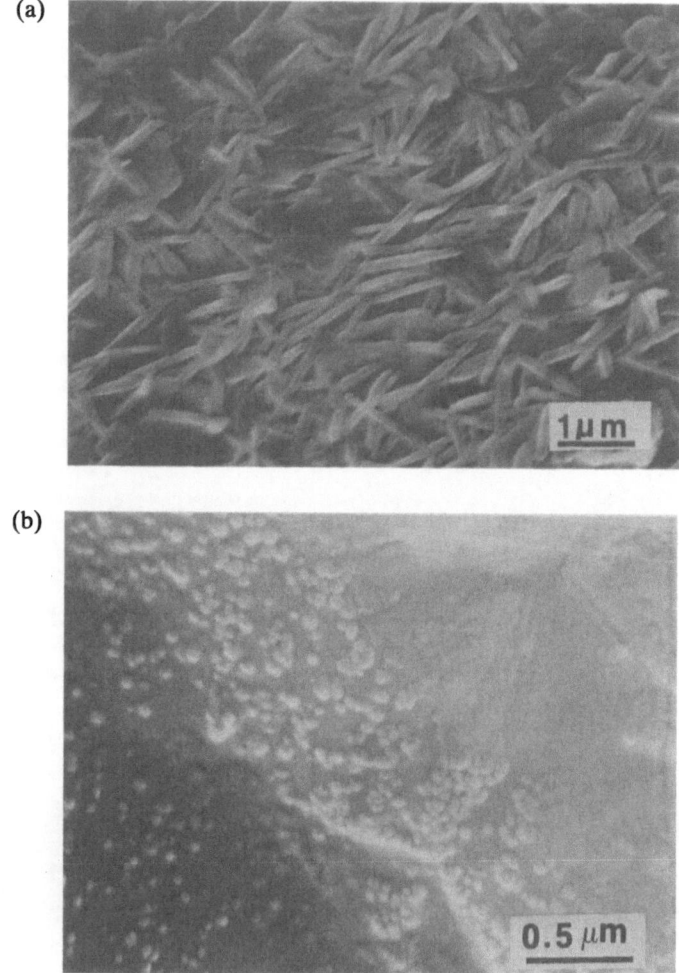

FIG. 5. (a) Effect of adding a structurally crystalline mullite to a di-phasic mullite gel and firing. (b) The original di-phasic gel fired alongside. The morphology of the final product has been changed dramatically.

TABLE 5
Technology-based experience with low-α materials

Based on three structures:		
Noncrystalline	SiO_2 glass (cf. high-quartz is slightly negative)	
Traditional	Zircon ($ZrSiO_4$)	
ceramics	Cordierite ($Mg_2Al_4Si_5O_{18}$)	Stuffed
New	Eucryptite–spodumene	derivatives
ceramics	($LiAlSiO_4$) ($LiAl_2Si_2O_6$)	of SiO_2

- Roy's arm-waving rule of *strongly bonded* (i.e. 3–4–5$^+$ ions) *chains* of polyhedra articulated around a *structural hole*
- Ternary or quaternary structure preferable to permit multiple-ion crystal chemical substitution

C. *Design nano- or micro-scale composites*
  - Assure thermodynamic compatibility of candidate pairs
  - Measure α of composite
  - Check for residual stresses

Since attempts to introduce other ions into eucryptite and spodumene had failed for 30 years we chose under strategy A, to introduce GeO_2 into cordierite. Others had tried this but failed because GeO_2 vaporized from a standard mixture. We applied the SSG technique and it worked beautifully. GeO_2-substituted cordierites proved to be excellent material with extremely low α [17].

Strategy B—finding a new structure—became possible owing to a serendipitous observation that French colleagues working on fast-ion conductors in the NASICON phase had increased their thermal shock resistance by a particular crystal chemical substitution. We also noted that this $NaZr_2P_3O_{12}$ structure fitted all the Roy guidelines for low α, and hence a thorough study was started.

In the event, an entire new class of totally controllable ultra-low-α materials was discovered. The crystal chemistry was extraordinary, indeed it was exactly the opposite of eucryptite since dozens of ionic substitutions were possible [18] and many of these gave us precise control of α in almost any range of interest [19]. Finally (Fig. 6), it even proved possible to eliminate the anisotropy of α_c and α_a in a $(CaSr)_{0.5}Zr_2P_3O_{12}$ phase, so that one had a zero-α, zero-anisotropy ceramic melting above 1700°C.

A further challenge was to make the material ferrimagnetic so that it would absorb microwave radiation. In this we failed. However, we

Rustum Roy

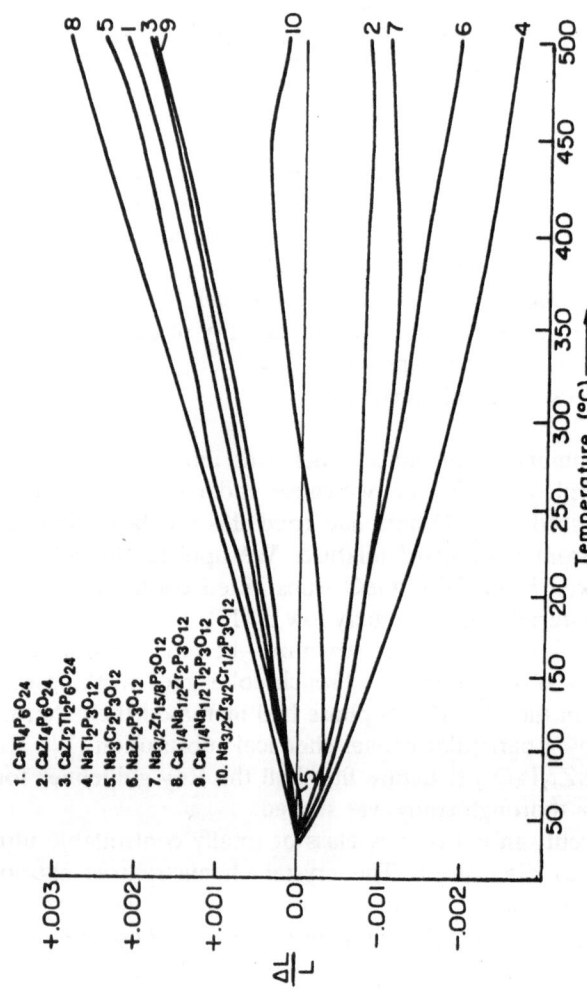

THERMAL EXPANSION OF [NZP]-FAMILY

1. $CaTi_4P_6O_{24}$
2. $CaZr_4P_6O_{24}$
3. $CaZr_2Ti_2P_6O_{24}$
4. $NaTi_2P_3O_{12}$
5. $Na_3Cr_2P_3O_{12}$
6. $NaZr_2P_3O_{12}$
7. $Na_{3/2}Zr_{15/8}P_3O_{12}$
8. $Ca_{1/4}Na_{1/2}Zr_2P_3O_{12}$
9. $Ca_{1/4}Na_{1/2}Ti_2P_3O_{12}$
10. $Na_{3/2}Zr_{3/2}Cr_{1/2}P_3O_{12}$

FIG. 6. The remarkable control possible of α value from negative to positive by changing the composition within the same (NZP) structure.

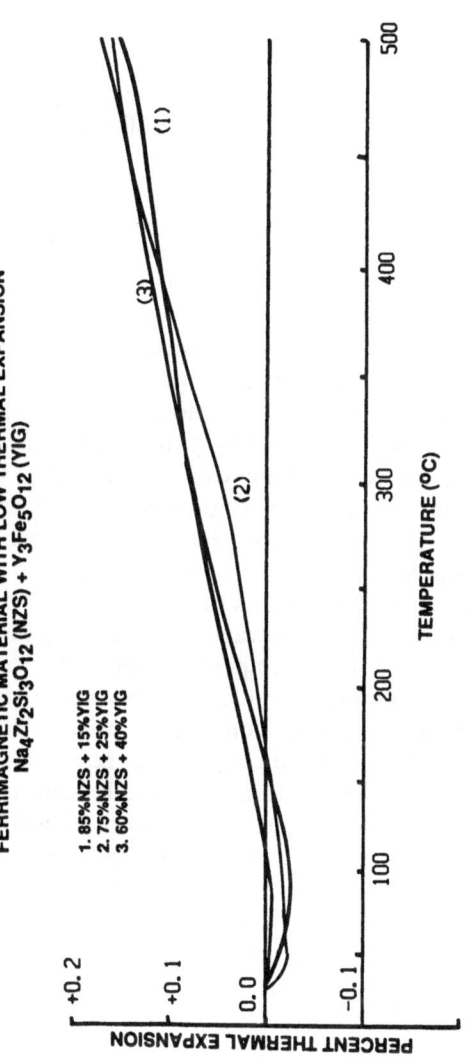

FIG. 7. No other strategy worked to make a ferrimagnetic zero-expansion ceramic for work in space environments. After determining that the two phases were stable together, a microcomposite was made.

also utilized strategy C—making a composite—to both make a zero-α composite and make it ferrimagnetic. Figure 7 shows the α vs T for the thermodynamically stable composite material consisting of YIG and a Na–Zr–P–Si phase.

This account is almost a textbook case of systematic materials design and synthesis. It should be clearly recognized that there was an element of chance in discovering the NZP family, but the other parts proceeded on a systematic path.

2.3. New Processes for Making Ceramic Powders

Fine ceramic powders made by decomposing the nitrates (or sulfates) of one-component oxides were commercialized for MgO and Al_2O_3 before the war. The sol–gel process was fully developed for making fine powders even in 5- and 6-component oxide systems in the early 1950s. The idea of decomposing sprays or mists on a high-temperature substrate was developed but not exploited at the same time.

In spite of intense interest in fine pure ceramic powders, virtually no new developments occurred for over two decades. Then the plasma methods and the laser decomposition methods appeared as new means to make ultra-pure sub-micrometer ceramic (typically non-oxide) powders. However, cost factors and inability to handle complex compositions seemed to have limited the potential of these methods. The opposite was true of the spray-tower method of Ruthner [20], who scaled up the process of decomposing a mist. We showed that oxides of virtually any composition [21] can be made directly into sub-micrometer particles at very competitive prices. However, although there was a long tradition and considerable research activity in making new powders in our laboratories, no really new method had been discovered anywhere in decades. The concept of making ceramic oxide powders by striking an arc under water while using non-noble metal electrodes was not related to any other current research. The background goes back to Faraday, Zsigmondy, and Svedberg.

In 1985 we devised the 'reactive electrode submerged arc' (RESA) process for making fine ceramic powders in a single apparatus (see Fig. 8 and Ref. 22). This process has proved to be a new general method for making fine sub-micrometer ceramic powders of Al_2O_3, TiO_2, ZrO_2, etc. Figure 9 shows uniform 100-nm Cr_2O_3 and ZnO powders made by this process. The former can be changed to Cr_3O_4 powder merely by substituting H_2O_2 for H_2O in the cell. SiC and TiC powders have been made by using non-aqueous fluids. More recently, binary

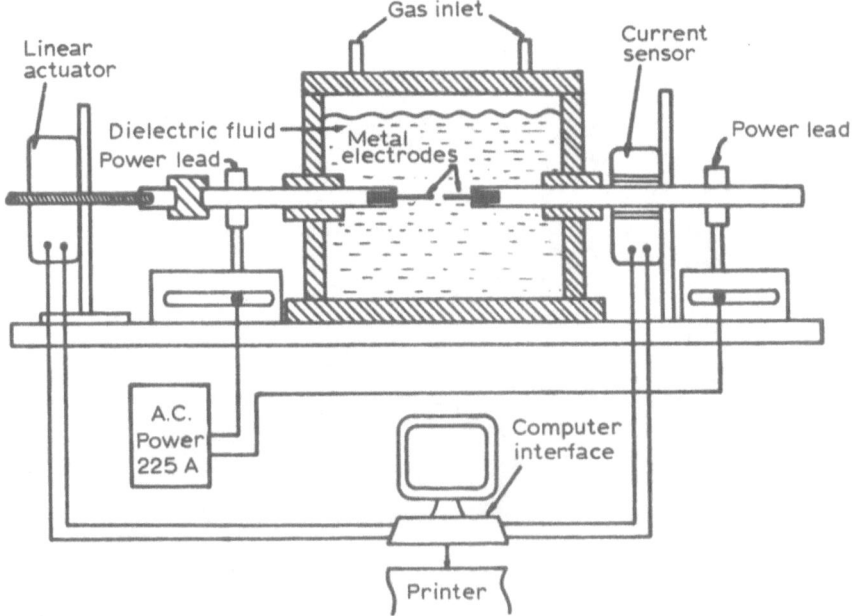

FIG. 8. Schematic of the newly designed RESA process.

oxide powders such as the spinels have been made. The process also yields a manageable sol of the fine powder. The research strategy here exemplifies the approach of high-risk innovation. It is easily measured by the number of papers in the field and the concurrent work.

2.4. The Derjaguin Laser High-pressure Phase Synthesis

As a follow-up on the research strategy noted above, a remarkable example may be found in one area of work of B. V. Derjaguin, the extraordinarily inventive and versatile Soviet chemist. Derjaguin, who had brought the world CVD diamond films in 1956–1976 [23], reported in 1982 that by merely exposing a falling stream of graphite particles in air to a CO_2 laser beam they could be converted into diamond. Except for his own follow-up paper in which he made quartz transform to stishovite, there was not a single follow-up study of this extraordinary claim.

In 1987 we followed our own strategy of selecting the highly unusual for study and attempted to confirm the work. We have indeed shown that CO_2 or Nd–YAG laser influences can effect high-pressure phase

(a)
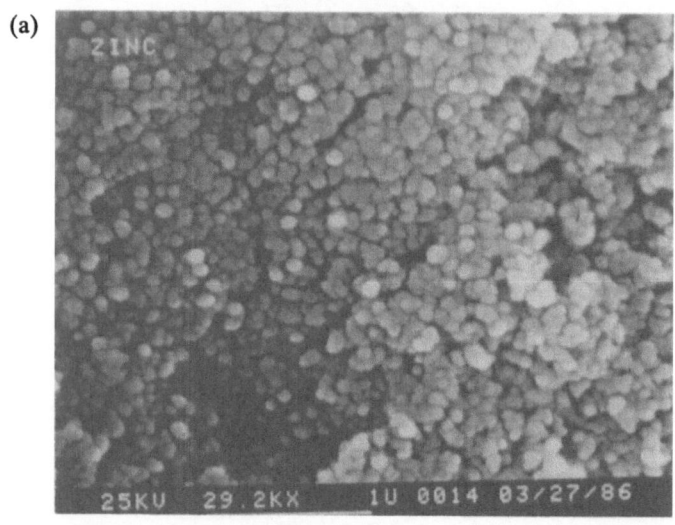

(b)

FIG. 9. Uniform, fine (100-nm) powders of (a) Cr_2O_3 and (b) ZnO produced by the RESA process.

synthesis in the laboratory; indeed we have succeeded in making even higher-pressure phases of SiO_2 than Derjaguin [24]. Because of our work, considerable interest has been sparked in the process, although we added but little to Derjaguin's. Initial skepticism at a truly revolutionary result appears to deter most investigators. This is probably the same reason why so few have worked on the Lanxide process referred to earlier.

2.5. High-T_c Superconductors

It was confirmed in the MRS Videotape History project on high-T_c oxides [25] that it was not the Bednorz and Muller paper that triggered the discovery of high T_c superconductivity but the rush was started to join the bandwagon after a paper was presented by Fueki at the MRS meeting on December 5. Yet that group in Tokyo did not increase the T_c but an experienced group confirmed that the 'LaBaCuO' phase was superconducting at 30 K. The announcement by Zhao in Peking of the YBaCu phase and confirmation soon by Wu in Alabama and Rao in Bangalore of T_c values near 90 K illustrated several important issues regarding research strategy (see Ref. 1 for details).

First: the element of chance. Raveau in France, Lazarev in Moscow and others had actually made the same phases years before but had simply not measured them to 4 K. Neither the Zhao group nor the Wu group had either experience or particular skill in materials design, synthesis or characterization: the finding of 1:2:3, although fortuitous to some extent, was guided by very elementary considerations of selecting similar ions to La^{3+}. Zhao wanted to avoid the f electron and chose Y. Wu argued (erroneously) from the pressure dependence data that smaller ions would be better. The fact that this family was so easily found, however, led to exactly the wrong conclusion by thousands of researchers world-wide that there must be many other phases waiting to be found. After some thousands of person-years of research only the Bi- and Tl-related structures have emerged. The consensus reports [2, 26] have cast doubt on much technological potential for these materials and most investigators find that almost anything they do has been done at least in part by someone else.

What of research strategy in such a situation? In my view it is exactly the opposite of what is being done in the United States at least. While it may be noted that relatively inexperienced groups 'first' synthesized these new phases, it could be also noted that it has taken the experienced groups a matter of days to a week or two to provide a clearer understanding of structure and composition of each of the key phases. In other words, the systematic empirical studies by experienced groups advocated in my strategy would inevitably have encountered the $1:2:3$ and the CaBiSrCu phases within roughly the same time frame. Several hundreds of person-years of research by less-experienced groups could have been saved. Many countries seem to be taking such an approach. It is certainly cost-effective and conserving of technical talent—the resource in shortest supply.

REFERENCES

1. R. Roy, HTSC: Restoring scientific and policy perspective. In *Proceedings of the World Congress on Superconductivity*, ed. C. G. Burnham. World Scientific, New Jersey, 1988, pp. 27–41.
2. J. H. Helm, Chair, USNAS, Committee on Science, Engineering and Public Policy, Panel on Superconductivity. NAS, Washington, DC, 1988.
3. A. Weinberg, *Minerva*, **I**, 159 (1963).
4. R. E. Newnham and L. E. Cross, *Mater. Res. Bull.*, **9**, 927 (1974).
5. M. L. Keith and R. Roy, *Am. Mineralogist*, **39**, 1–23 (1959).
6. R. Roy, *Science*, **238**, 1664 (1987).
7. R. C. Garvie, R. H. Hannink and R. T. Pascoe, *Nature*, **258**, 703 (1975).
8. T. C. Simonton, R. Roy, S. Komarneni and E. Breval, *J. Mater. Res.*, **1**, 667 (1986).
9. R. A. Roy and R. Roy, *Mater. Res. Bull.*, **19**, 169 (1984). D. Hoffmann, S. Komarneni and R. Roy, *J. Mater. Sci. Lett.*, **3**, 439 (1984). D. Hoffmann, R. Roy and S. Komarneni, *Mater. Lett.*, **2**, 245 (1984). *J. Am. Ceram. Soc.*, **67**, 468 (1984).
10. G. Ervin, Ph.D. Thesis, Pennsylvania State College (1949).
11. A. Kijowski, S. Komarneni and R. Roy, Effect of seeding on the crystallization of cordierite. *In 90th Fall Meeting of Am. Ceram. Soc.*, Abstracts (1988).
12. D. M. Roy and R. Roy, *Nat. Res. Council Publ.* **456**, 82 (1956).
13. R. Roy, Y. Suwa and S. Komarneni, *Nucleation and epitaxial growth in diphasic crystalline + amorphous gels*. In *Science of Ceramic Chemical Processing*. ed. L. L. Hench and R. R. Ulrich, Wiley, New York, 1986, Vol. 2, Chap. 27.
14. F. A. Hummel, *Foote Prints*, **20**, 3 (1948).
15. R. Roy and E. F. Osborn, *J. Am. Chem. Soc.*, **71**, 2086 (1949).
16. L. G. van Uitert, H. M. O'Bryan, M. E. Lines, H. J. Guggenheim and G. Zydzik, *Mater. Res. Bull.* **12**, 261 (1977).
17. D. K. Agrawal, V. S. Stubican and Y. Mehrotra, *J. Am. Ceram. Soc.*, **69**, 261 (1977).
18. J. Alomo and R. Roy, *J. Mater. Sci.*, **21**, 444 (1986).

19. R. Roy, D. K. Agrawal and R. A. Roy, United States Patent 4,675,302 (June 23, 1987).
20. M. J. Ruthner, Preparation and sintering characteristics of MgO, MgO–Cr_2O_3 and MgO–Al_2O_3, Third Round Table Meeting, Intl. Team for Studying Sintering, Herceg-Novi, Yugoslavia, 3–8 Sept. 1973.
21. D. M. Roy, R. R. Neurgaonkar, T. P. O'Holleran and R. Roy, *Ceramic Bull.*, **56**, 1023 (1977).
22. A. Kumar and R. Roy, RESA—A wholly new process for fine oxide powder preparation, *J. Mater. Res.* **3**, 1373 (1988).
23. B. V. Derjaguin *et al.*, Filamentary diamond crystals. *J. Cryst. Growth*, **2**, 380 (1968).
24. M. Alam, T. Debroy and R. Roy, Diamond formation in air by the Fedoseev–Derjaguin laser process, *Carbon*, **27**, 289 (1989).
25. *History of High-T_c Superconductors*, Materials Research Society History Project, videotapes.
26. Journal Electric Power Research Institute; p. 23, Dec. 1987.

2

Equilibrium Relations Involving Transition-Metal Oxides at High Temperatures

ARNULF MUAN

Department of Geosciences, The Pennsylvania State University
University Park, Pennsylvania, USA

ABSTRACT

Equilibrium relations in oxide systems involving transition metals as components are discussed in terms of phase diagrams and thermodynamic properties of individual solid-solution phases. It is shown that at high temperatures ($\sim 1400°$–$1700°C$) and very low oxygen pressures ($P_{O_2} \sim 10^{-10} - 10^{-13}$ atm) the constituent transition-metal ions occur in unusual oxidation states and that, consequently, the chemistry and the phase relations under these conditions differ dramatically from those prevailing in air. Systems containing chromium oxide or titanium oxide are used as examples, and some general conclusions are drawn. Finally, projections of future directions of research in this area are presented.

1. INTRODUCTION

Materials science and geosciences encompass systems and processes in which complex heterogeneous reactions take place and involve the coexistence of a number of crystalline and liquid phases, both metals and non-metals, e.g. oxides, sulfides, carbides and nitrides, at high temperatures. Furthermore, these condensed phases (i.e. crystalline and liquid) have important interactions with components of gas phases, especially oxygen. Predictions and control of the behavior and properties of such systems at high temperatures require quantitative knowledge of the chemical properties of all phases present, including the gas phase. This is particularly so in systems where transition-metal oxides are present, because these oxides typically involve ions in several different oxidation states, and their relative amounts, and hence the phase relations, are strongly dependent on the oxygen partial pressure under which the reactions take place. The present paper gives a review of such relations and makes some projections into the future.

Only equilibrium aspects of such systems will be considered in the present paper. At high temperatures a sufficiently close approach to equilibrium is commonly realized to make such data extremely important, although it must be recognized that in some instances kinetic aspects may be of equal importance.

A few examples of completed work or work in progress will be selected in order to demonstrate the principles involved, to emphasize general features, and to draw some general conclusions that may serve as guidelines for projections into the future. The presentation will be divided into two main parts, viz., (1) phase relations in some oxide systems at relatively high temperatures (1400–1800°C) and strongly reducing conditions ($P_{O_2} < 10^{-10}$ atm), and (2) thermodynamic aspects of individual oxide solid solutions.

2. PHASE RELATIONS INVOLVING SELECTED TRANSITION-METAL OXIDES

By far the most extensively studied systems involving transition-metal oxides as components are those containing iron oxides. The methodology for studying and representing such systems in legible form was established in three main stages in the 1930s [1], 1940s [2, 3] and 1950s [4, 5]. It has become increasingly clear in more recent years that similar factors play essential roles in a large number of additional transition-metal oxide systems at high temperatures, and that indeed the relations in some cases may be more complex than in the iron oxide-containing systems because more than two different oxidation states may be present in significant amounts. In the present case it seems appropriate to use chromium oxide- and titanium oxide-containing systems as examples, partly because they are technologically important, partly because the honoree of this symposium, Shigeyuki Sōmiya, got his early start in oxide chemistry working with the author on Cr_2O_3-containing systems, and other participants at this

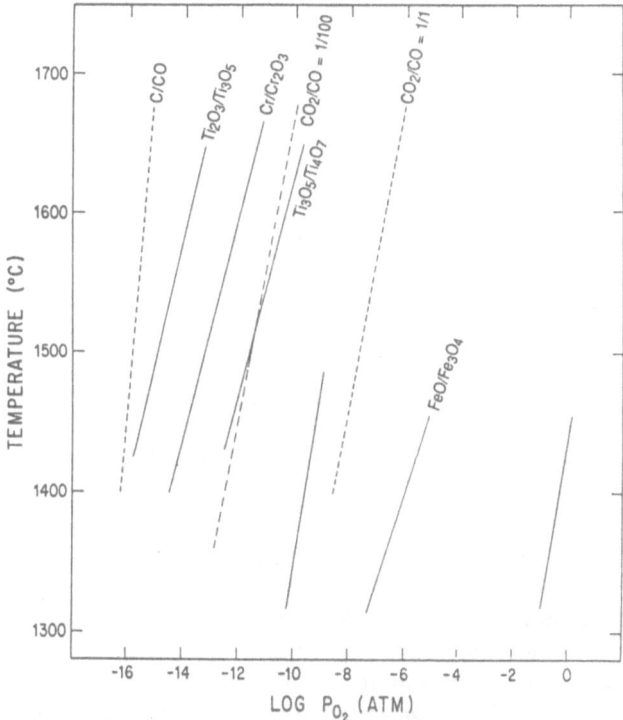

FIG. 1. Univariant equilibria of selected oxide/metal pairs, as well as the CO/C equilibrium, and CO_2/CO ratios of a coexisting gas phase, as a function of temperature and oxygen partial pressure at a total pressure of 1 atm.

symposium worked actively on various aspects of titanium oxide-containing systems at that time.

What I shall discuss in the present paper, however, differs markedly from what we did on those systems previously. Some of the differences are illustrated with reference to the diagram in Fig. 1, showing temperature–oxygen pressure relations for selected univariant equilibria. It is seen that the equilibria involving coexistence of Cr_2O_3 with metallic Cr and the stability ranges of the titanium oxides Ti_2O_3 and Ti_3O_5 occur at oxygen pressures many orders of magnitude below those involved in the case of the equilibrium between FeO and metallic Fe or between Fe_3O_4 and Fe_2O_3. Hence, different and more demanding experimental procedures must be applied in the case of the Cr_2O_3/Cr and the Ti_2O_3/Ti_3O_5 containing systems. In practice,

CO_2/CO (or CO_2/H_2) gas mixing ratios below 1/100 are difficult to control accurately. Another difficulty is the possible precipitation of carbon by decomposition of CO, or the formation of carbide or oxycarbide phases if CO_2 or CO-containing gases are used to control the oxygen pressure under strongly reducing conditions. Dashed curves representing the C/CO equilibrium, as well as CO_2/CO at various mixing ratios of the two gases, are shown in Fig. 1. It is also to be noted that the temperature ranges used in the more recent studies of the chromium oxide- and the titanium oxide-containing systems in general are several hundred degrees higher than those commonly used in most of the work on iron oxide-containing systems. Another difference between the early and more recent work is that the early work was focused on 'mapping' of phase diagrams, whereas in the present we are focusing on determination and evaluation of thermo-dynamic properties (activity–composition relations) of individual phases. These latter aspects of the phase diagrams will be discussed in Chapter 3.

Examples of the effects of varying oxygen pressures on equilibrium relations in chromium oxide-containing systems are shown in Figs 2–4, in order to demonstrate that the chemistry of these systems is entirely different under strongly reducing conditions. Figure 2 shows phase relations in the system chromium oxide–SiO_2 at two different levels of oxygen pressure. The diagram on the left [6, 7] (Fig. 2(a)) applies when the atmosphere is air ($P_{O_2} = 0 \cdot 2$ atm) and the diagram on the right (Fig. 2(b)) under conditions sufficiently reducing (contact with metallic Cr [8]) to maintain most of the chromium in the divalent state. It is seen that under these conditions the liquidus (and solidus) temperatures have been drastically lowered, as compared to those prevailing in air, and also that the mutual solubilities of the two oxides in the liquid state have increased drastically. Furthermore, silica (cristobalite or tridymite, depending on temperature) is the crystalline phase in equilibrium with the two liquids rather than eskolaite (Cr_2O_3), as was the case in air.

Similar effects are observed in more complex chromium oxide-containing silicate systems. Examples for 'ternary' systems† are shown

---

† These systems are not strictly ternary, because chromium is present in different oxidation states in varying amounts, depending on temperature and basicity of the liquids. However, the systems may be portrayed in triangular diagrams for sake of simplicity.

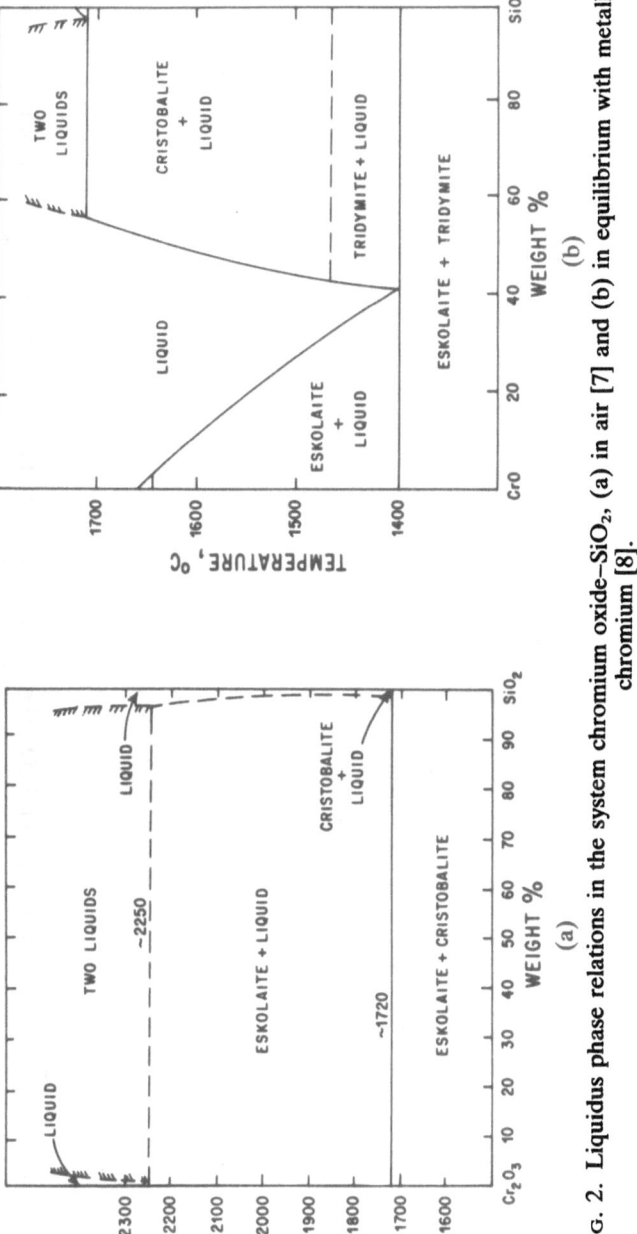

FIG. 2. Liquidus phase relations in the system chromium oxide–SiO₂, (a) in air [7] and (b) in equilibrium with metallic chromium [8].

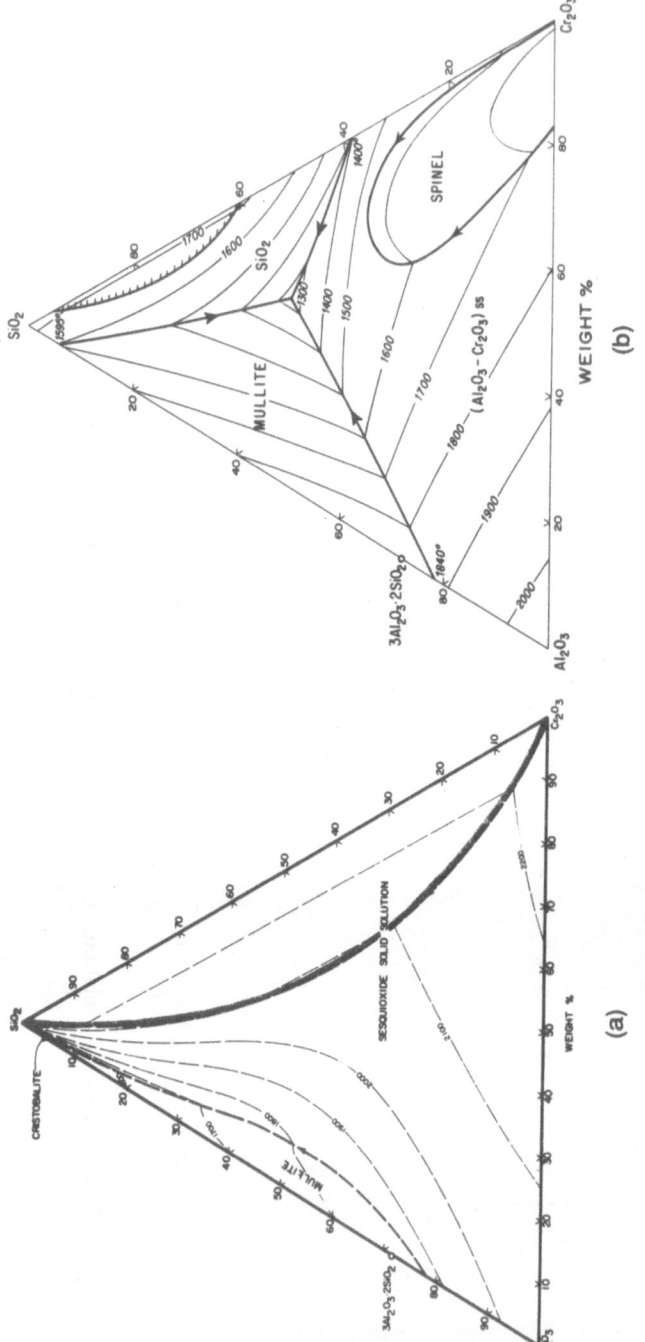

Fig. 3. Liquidus phase relations in the system Al₂O₃–chromium oxide–SiO₂, (a) in air [9] and (b) in equilibrium with metallic chromium [10].

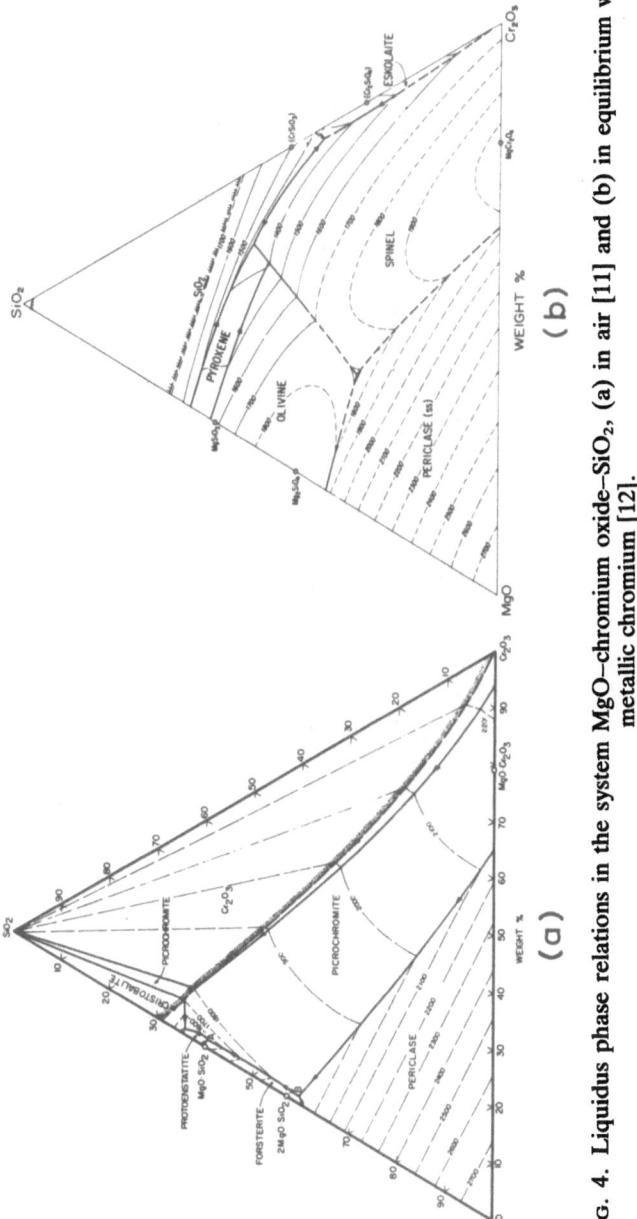

Fig. 4. Liquidus phase relations in the system MgO–chromium oxide–SiO₂, (a) in air [11] and (b) in equilibrium with metallic chromium [12].

in Figs 3 and 4, showing phase relations in the systems Al_2O_3–chromium oxide–SiO_2 and MgO–chromium oxide–SiO_2, respectively.

In both of these systems, the effect of lowering the oxygen pressure is to lower liquidus and solidus temperatures, increase the solubility of chromium oxide in the silicate liquid, and increase the solubility of 'CrO' in the crystalline phases when a suitable host for Cr^{2+} is present, e.g. in periclase, picrochromite, forsterite and protoenstatite in the system MgO–chromium oxide–SiO_2 in contact with metallic Cr. Similar effects have been observed in the system CaO–chromium oxide–SiO_2 in contact with metallic chromium [13]. In the latter system a garnet phase, $Ca_3Cr_2Si_3O_{12}$, is stable on the liquidus surface.

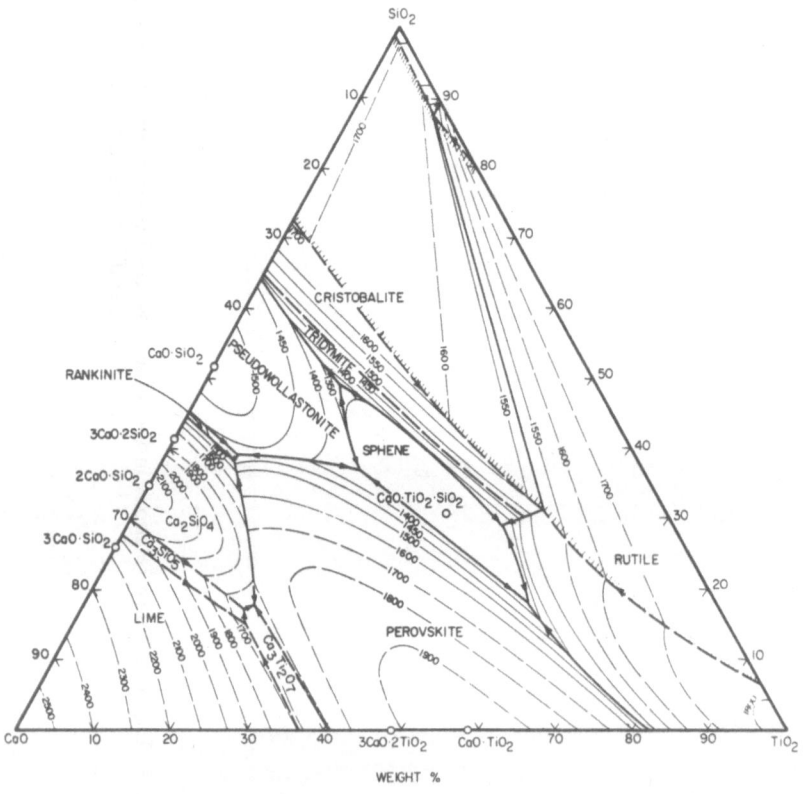

FIG. 5. liquidus phase relations in the system CaO–titanium oxide–SiO_2 in air [16].

The appearance of this phase at liquidus temperatures under strongly reducing conditions, as opposed to its absence as a stable crystalline phase on the liquidus surface of the analogous system under more oxidizing conditions [14, 15], is a result of the lowering of liquidus temperatures under strongly reducing conditions. Whereas the garnet phase has essentially the same composition in the two cases, it decomposes in air to the phase assemblage pseudowollastonite plus eskolaite at temperatures below the solidus.

The effects of strongly reducing conditions on phase relations in titanium oxide-containing silicate systems are also significant but somewhat different from those observed in the chromium oxide-containing silicate systems just discussed. For instance, in the system CaO–titanium oxide–SiO$_2$ shown in Fig. 5 and 6, the effect on liquidus and solidus temperatures of lowering the oxygen pressure is relatively

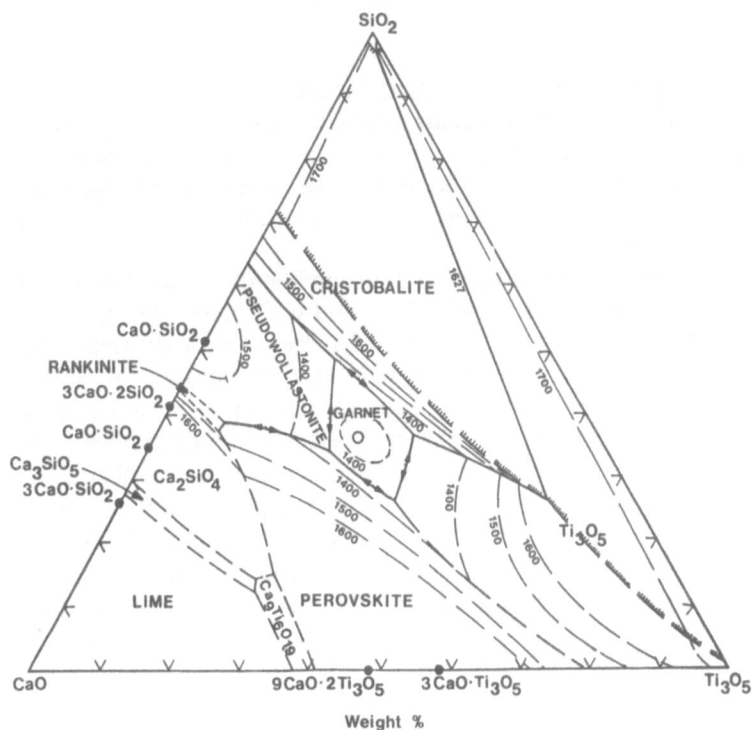

Fig. 6. Liquidus phase relations in the system CaO–Ti$_2$O$_3$–SiO$_2$ [17].

small. However, a new ternary crystalline phase of garnet structure replaces sphene on the liquidus surface when the conditions become sufficiently reducing to stabilize Ti^{3+}. The composition of this phase is approximately $Ca_3Ti_2Si_3O_{12}$ although it is somewhat off stoichiometry. Based on our recent work [18], we can also portray the main features of phase relations within the quaternary system $CaO-Ti_2O_3-TiO_2-SiO_2$, as will be reported in detail elsewhere. It is noteworthy that extensive liquid miscibility gaps exist in the joins $Ti_2O_3-SiO_2$ and $Ti_3O_5-SiO_2$ and that no intermediate crystalline phase analogous to mullite is stable in these systems. Indeed, the behavior of Ti_2O_3 in silicate systems seems to be much more similar to that of Cr_2O_3 than that of Al_2O_3, in contradiction to inferences made some 30 years ago [19].

3. ACTIVITY–COMPOSITION RELATIONS IN OXIDE SOLID SOLUTIONS

While temperature–composition phase diagrams, examples of which were presented in the foregoing, provide much useful, practical information on stability ranges of the various phases appearing in these diagrams, the ultimate goal is to understand quantitatively the thermodynamic parameters governing heterogeneous equilibria and heterogeneous reactions.

Activity–composition relations for a number of moderately stable transition-metal oxides, e.g. FeO, CoO and NiO, have been available for some time, whereas similar data for more stable transition-metal oxide components have been largely lacking because of severe experimental difficulties attending such determinations. Among these difficulties are the very low oxygen pressures associated with the coexistence of alloy and oxide phases involving the more stable oxides, and hence the problem of controlling the oxygen pressures accurately by use of common gas-mixing techniques, and the relatively high temperatures (1500–1800°C) commonly associated with these equilibria. In some cases, vaporization losses may be appreciable under these circumstances, and special care must be taken to minimize or account for the effects of such factors.

A key to derivation of activity–composition relations of relatively stable oxides (e.g. chromium oxide and titanium oxide) has been the determination, in our laboratories, of activity–composition relations in some key alloy systems, viz. Mo–Cr, Pt–Cr and Pt–Ti. Activity–composition relations in these alloys were calculated from equilibrium measurements involving molybdenum or platinum metal, chromium oxide or titanium oxide of unit activity, and H_2–CO_2 gas mixtures of known oxygen partial pressures. Both these metals (Mo and Pt) can be used to relatively high temperatures ($\sim$1600–1800°C), depending on the nature of the oxide equilibrium to be studied. Activity–composition relations in Mo–Cr and Pt–Cr alloys, as determined in our laboratories [20, 21], are shown in Figs 7 and 8, respectively. The Pt alloy above has strong negative deviation from ideality. The negative deviation from ideality is even more pronounced in the system Pt–Ti, the limiting activity coefficient of which at 1500°C is $\sim$10^{-9} [22]. These strong negative deviations permit determination of activity–composition relations for relatively stable oxide components such as chromium oxide and titanium oxide at oxygen pressures that are sufficiently high to permit their accurate control by use of CO_2–H_2 gas mixtures, and preservation of the transition-metal oxide in its most stable oxidation state, viz., Cr^{3+} and Ti^{4+} for chromium and titanium, respectively.

The underlying equations for determination of oxide activities from

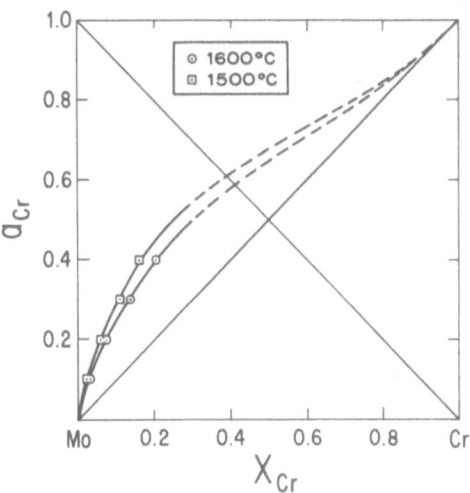

Fig. 7. Activity–composition relations in Mo–Cr alloys at 1500°C and 1600°C [20].

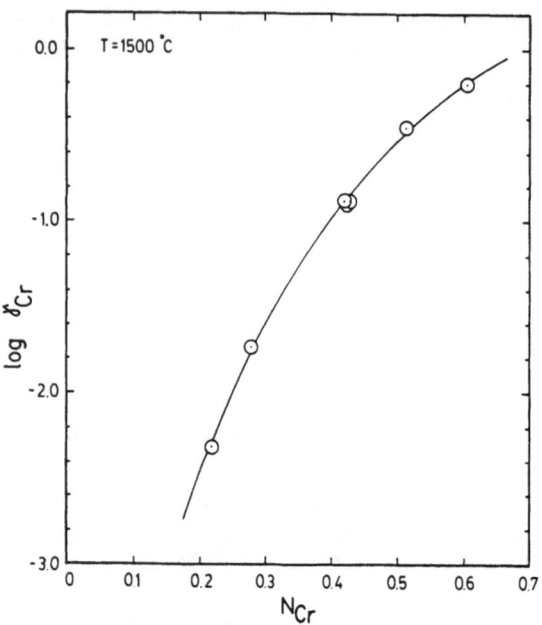

FIG. 8. Activity–composition relations in Pt–Cr alloys at 1500°C [21].

oxide–alloy–gas equilibrations are as follows. Consider the coexistence of oxide, alloy and gas according to the reaction

$$[\text{Me}] + \tfrac{1}{2}O_2(g) = (\text{MeO}) \tag{1}$$

where square brackets indicate the element Me in an alloy phase and the parentheses indicate the conjugate oxide component in an oxide solution. The equilibrium can be expressed by the equation

$$K = \frac{a_{\text{MeO}}}{a_{\text{Me}} \cdot P_{O_2}^{1/2}} \tag{2}$$

where K is related to the standard Gibbs free energy change by the relation $\Delta G^0 = -RT \ln K$. From eqn (2) the activity of the oxide may then be calculated as

$$a_{\text{MeO}} = K \cdot a_{\text{Me}} \cdot P_{O_2}^{1/2} = \left(\frac{P_{O_2}}{P_{O_2}^0}\right)^{1/2} \tag{3}$$

where $P_{O_2}^0$ is the oxygen pressure for the coexistence of pure metal

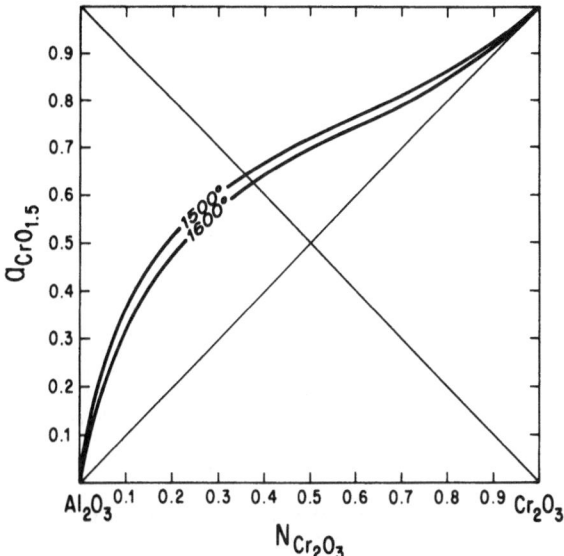

Fig. 9. Activity–composition relations in Al_2O_3–Cr_2O_3 solid solutions at 1500 and 1600°C [23].

(Me) and pure oxide (MeO). An example of results obtained is presented in Fig. 9, showing activity–composition relations for Al_2O_3–Cr_2O_3 solid solutions at 1500 and 1600°C [23]. Based on these data, activity–composition relations for $FeAl_2O_4$–$FeCr_2O_4$ solid solutions at 1500°C (see Fig. 10) have been calculated from experimentally determined directions of conjugation lines between coexisting sesquioxide and spinel solid solutions in the system FeO–Al_2O_3–Cr_2O_3 [23]. The theoretical basis for deriving thermodynamic relations from such conjugation lines was presented by the author more than 20 years ago [24].

In the thermodynamic treatment of equilibria involving multicomponent phases with more complicated structures, special caution must be exercised in selecting the most appropriate activity expressions for the solid solutions involved. The general relationships governing such cases have been treated by statistical-mechanical methods by a large number of authors [25, 26]. For a solid solution $A_uB_vZ_w$–$C_uD_vZ_w$ ($=(A, C)_u^\alpha(B, C)_v^\beta Z_w$), in which there are two distinctly different cation

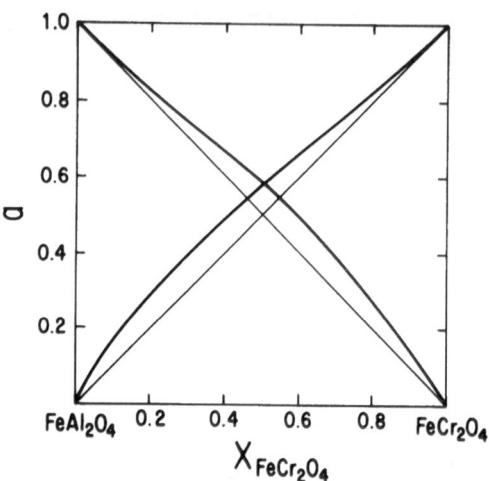

FIG. 10. Activity–composition relations in $FeAl_2O_4$–$FeCr_2O_4$ solid solutions in equilibrium with metallic iron at 1500°C [23].

sites (α and β), the ideal activity expressions of the two end-member components are

$$a_{A_uB_vZ_w} = K \cdot (X_A)^u (X_B)^v \qquad (4)$$

and

$$a_{C_uD_vZ_w} = K \cdot (X_C)^u (X_D)^v \qquad (5)$$

respectively, where K is a constant (equal to 1 if the end-member components are used as reference states), and x with appropriate subscripts is the atom fraction of A and C, or of B and D, on the α- and β-sites of the structure, respectively. For a solution, $(A, B)_u Z_w$ involving mixing on only one atomic site, the equations above become

$$a_{A_uZ_w} = (X_{A_uZ_w})^u \qquad (6)$$

and

$$a_{B_uZ_w} = (X_{B_uZ_w})^u \qquad (7)$$

Applied to an ideal sesquioxide solid solution, for instance Al_2O_3–Cr_2O_3, the expressions would be

$$a_{Cr_2O_3} = X^2_{Cr_2O_3} \qquad (8)$$

or

$$a_{CrO_{1.5}} = X_{Cr_2O_3} \qquad (9)$$

The activities shown in Fig. 9 are in accordance with the expression in the latter equation.

Examples of activity–composition relations in titanate solid solutions are shown in Figs 11 and 12 for the metatitanate join in the system $MgO–CoO–TiO_2$ [27] and $MnO–CoO–TiO_2$ [28], respectively. In the former system (Fig. 11) the activity–composition relations are close to ideal, because the solid solution involves the substitution of two cations (Mg^{2+} and Co^{2+}) of approximately equal size on one cation sublattice, whereas in the latter system (Fig. 12) there is a

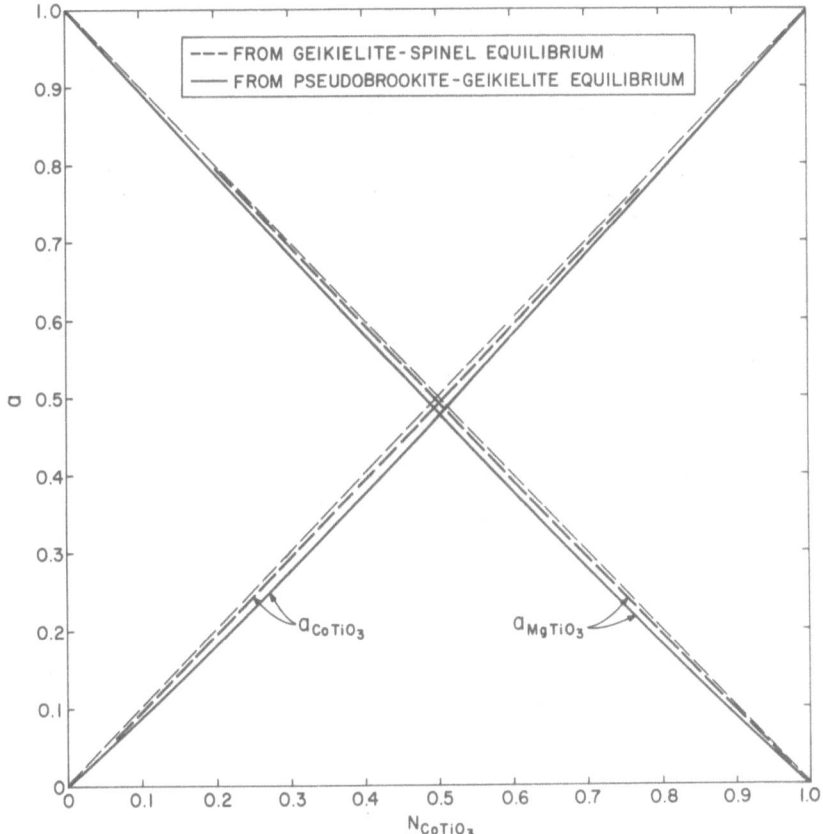

FIG. 11. Activity–composition relations along the metatitanate join in the system $MgO–CoO–TiO_2$ at 1300°C [27].

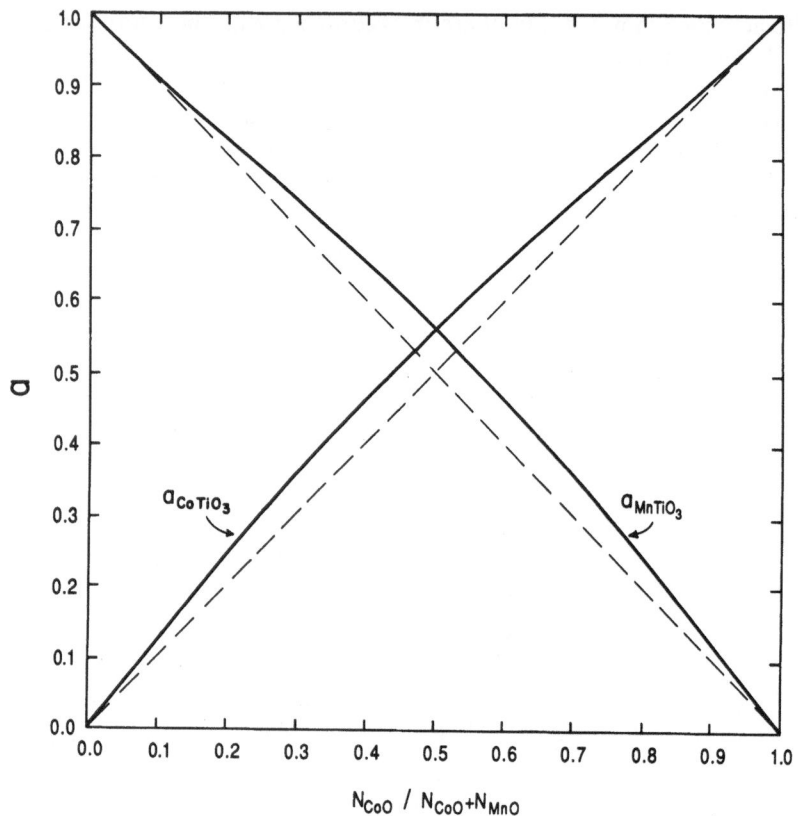

Fig. 12. Activity–composition relations along the metatitanate join in the system MnO–CoO–TiO₂ at 1250°C [28].

moderate positive deviation from ideality because of the larger size difference between the two substituting cations (Mn^{2+} and Co^{2+}).

Determinations of activity–composition relations of chromium oxide and titanium oxide in silicate liquids have also been carried out recently in our laboratories by methods similar to those described for solid solutions in the foregoing. The data for the liquid phase are still being evaluated and will be published elsewhere in the near future.

4. PRESENT STATUS AND FUTURE PROJECTIONS

Much progress has been made during the past decade towards quantitative understanding of the thermodynamic parameters governing reactions in and among oxide phases at high temperatures, mainly in the crystalline state. This progress includes both experimental techniques and theoretical treatment of such equilibria. In the former area the recently acquired knowledge of activity–composition in some key alloy systems (e.g. Mo–Cr, Pt–Cr, Pt–Ti) at high temperatures has been instrumental in expanding the regime of accurate thermo-dynamic studies of oxide phases to higher temperatures and lower oxygen pressures than was possible previously. In the second category (theory) the increased application of statistical mechanics and model-ing based on realistic assessment of structural characteristics of the solutions have permitted the formulation of reasonable and useful generalizations from which interpolations and extrapolations to temperature–pressure–composition regions for which experimental data are lacking may be made with greater assurance than was possible previously. Furthermore, the increased database and greatly expanded understanding of the thermodynamic parameters governing high-temperature reactions among oxides in general and transition-metal oxides in particular provide a valuable base for deriving a better understanding also of the kinetics and mechanisms of oxide reactions at high temperatures.

What, then, are some of the most needed or most promising new directions in our quest to learn more about the chemical behavior of transition-metal oxide phases and their solutions at high temperatures?

First, there is a tremendous potential for synthesizing new phases or incorporating transition metal ions of unusual oxidation states into known, stable host structures of various symmetries, by judicious control of oxygen pressure, chemical composition and temperature, in regimes of these parameters which have not been explored previously. The development of new alloys and other materials for containing some of these new phases at high temperatures will also be a crucial factor for success in such experiments and syntheses. While our understanding of many crystalline solid solutions is reasonably good, a similarly systematic treatment and universally valid generalization for liquids is not so clearly available. New experimental data becoming available in our laboratories and elsewhere, in combination with computer modeling work, hold promise of considerable progress in this field during the next few years. It should be emphasized, however, that modeling and carefully selected experimental verification with reliable 'calibration points' must go hand in hand in order to establish a sound basis for accurate thermodynamic treatment of these complex phases. It also appears that more quantitative information must be obtained on deviations from stoichiometry and the effects of such deviations on chemical potentials, bonding characteristics and vacancy–ion interactions in crystalline solutions.

In dealing with transition-metal oxides with large vacancy concentrations it is probably unrealistic to assume that the structural characteristics existing at high temperatures can be preserved to room temperature by even the fastest quenching of the samples. It appears that we must increasingly move toward in-situ examination of such phases. New techniques hold great promise of shedding new light on such problems from the standpoint of both equilibrium and kinetic aspects of the behavior of such phases at high temperatures.

Much attention in high-temperature oxide chemistry has been directed towards substitutions on the cation sublattices, while relatively little attention has been devoted to substitutions on the anion sublattices. There seem to be unexplored possibilities in the latter area. In the context of the present paper, attention is directed to the relative closeness in Fig. 1 of the various oxide–metal univariant curves (e.g. Cr_2O_3–Cr) to the univariant curve of C/CO (at 1 atm pressure). Hence, we are in the regime where carbides and oxycarbides may form. Many of the latter phases will probably have desirable physical and mechanical properties as well as offering some unique bonding characteristics.

Finally, it is emphasized that better bridge-building must take place between those of us who study the chemistry of oxide phases at high temperatures and those who study physical and mechanical properties. Furthermore, we are all dependent on structural information of the highest possible resolution. In other words, we must use an interdisciplinary approach. The broad spectrum of interests represented by the participants at the Shigeyuki Sōmiya symposium is a good example of interaction across disciplinary borders.

ACKNOWLEDGMENTS

Data and ideas presented in this paper were derived mainly in conjunction with research projects sponsored by the American Iron and Steel Institute, the US Department of Energy, the Naval Sea Systems Command, and the Applied Research Laboratory Exploratory and Foundational Research Program.

REFERENCES

1. N. L. Bowen and J. F. Schairer, The system FeO–SiO$_2$, *Am. J. Sci.*, **24**, 177–213 (1932).
2. L. S. Darken and R. W. Gurry, The system iron–oxygen. I. The wüstite field and related equilibria, *J. Am. Chem. Soc.*, **67**, 1398–412 (1945).
3. L. S. Darken and R. W. Gurry, The system iron–oxygen II. Equilibrium and thermodynamics of liquid oxide and other phases, *J. Am. Chem. Soc.*, **68**, 798–816 (1946).
4. A. Muan, Phase equilibria in the system FeO–Fe$_2$O$_3$–SiO$_2$, *Trans. AIME*, **203**, 965–76 (1955).
5. A. Muan, Phase equilibria at high temperatures in oxide systems involving changes in oxidation states, *Am. J. Sci.*, **256**, 171–207 (1958).
6. E. N. Bunting, Phase equilibria in the system Cr$_2$O$_3$–SiO$_2$, *J. Res. Nat. Bur. Standards*, **5**, 325–7 (1930); RP203.
7. A. Muan and E. F. Osborn, *Phase Equilibria Among Oxides in Steelmaking*, Addison-Wesley Publishing Co., Inc., Reading, Mass., 1965.
8. H. Collins and A. Muan, Phase relations in the system chromium oxide–SiO$_2$ in equilibrium with metallic chromium, in preparation.
9. P. L. Roeder, F. P. Glasser and E. F. Osborn, The system Al$_2$O$_3$–Cr$_2$O$_3$–SiO$_2$, *J. Am. Ceram. Soc.*, **51**(10) 585–94 (1964).
10. K. Kitayama, E. Pretorius and A. Muan, Equilibrium relations in the system Al$_2$O$_3$–chromium oxide–SiO$_2$ under strongly reducing conditions, in preparation.
11. M. L. Keith, Phase equilibria in the system MgO–Cr$_2$O$_3$–SiO$_2$, *J. Am. Ceram. Soc.*, **37**, 490–6 (1954).
12. H. Collins, G. Merkel and A. Muan, Equilibrium relations in the system MgO–chromium oxide–SiO$_2$ under strongly reducing conditions, in preparation.
13. J. P. R. DeVilliers, J. Mathias and A. Muan, Phase relations in the system CaO–chromium oxide–SiO$_2$ in air and solid-solution relations along the Ca$_2$SiO$_4$–Ca$_3$(CrO$_4$)$_2$ join, *Trans. Inst. Mining and Metal. Sec. 3, Min. Proc. and Extr. Metal*, **96**, C55–C62 (1987).
14. F. P. Glasser and E. F. Osborn, Phase equilibrium studies in the system CaO–Cr$_2$O$_3$–SiO$_2$, *J. Am. Ceram. Soc.*, **41**, 358–67 (1958).
15. J. P. R. DeVilliers and A. Muan, Equilibrium relations in the system CaO–chromium oxide–SiO$_2$ under strongly reducing conditions, in preparation.
16. R. C. DeVries, R. Roy and E. F. Osborn, Phase equilibria in the system CaO–TiO$_2$–SiO$_2$, *J. Am. Ceram. Soc.*, **38**, 158–71 (1955).
17. V. J. Bruni and A. Muan, Phase relations in the system CaO–Ti$_2$O$_3$–SiO$_2$, in preparation.
18. V. J. Bruni and A. Muan, Phase relations in the system CaO–Ti$_2$O$_3$–TiO$_2$–SiO$_2$, in preparation.
19. R. Roy, R. C. DeVries, D. E. Rase, M. W. Shafer and E. F. Osborn, System Ti$_2$O$_3$–SiO$_2$–TiO$_2$; not impossible form, *Quarterly Report on Contract Nc. DA36-039*, sc-5594, Pennsylvania State College, 1959.

20. R. Snellgrove, D. Sain, T. Tsai, L. S. Darken and A. Muan, Activity–composition relations in Mo–Cr alloys at 1500 and 1600°C, in preparation.
21. E. Pretorius and A. Muan, Activity–composition relations in Pt–Cr alloys at 1500°C, in preparation.
22. V. J. Bruni and A. Muan, Activity–composition relations in Pt–Ti alloys at 1500°C, in preparation.
23. T. Tsai and A. Muan, Experimental determination of activity–composition relations in Al_2O_3–Cr_2O_3 solid solutions at 1500 and 1600°C, in preparation.
24. A. Muan, Determination of thermodynamic properties of silicates from locations of conjugation lines in ternary systems, *Am. Mineral.*, **52**, 797–804 (1967).
25. K. Schwerdtfeger and A. Muan, Activities in olivine and pyroxenoid solid solutions in the system Fe–Mn–Si–O at 1150°C, *Trans. AIME*, **236**, 201–11 (1966).
26. D. M. Kerrick and L. S. Darken, Statistical thermodynamic models for ideal oxide and silicate solid solutions, with application to plagioclase, *Geochim. Cosmochim. Acta*, **39**, 1431–42 (1975).
27. B. Brezny and A. Muan, Activity–composition relations of solid solutions and stabilities of Mg_2TiO_4, $MgTiO_3$ and $MgTi_2O_3$ as determined from equilibria in the system MgO–CoO–TiO_2 at 1300°C, *Thermochim. Acta*, **2**, 107–19 (1971).
28. L. G. Evans and A. Muan, Activity–composition relations of solid solutions and stabilities of the manganese and nickel titanates at 1250°C as derived from equilibria in the systems MnO–CoO–TiO_2 and MnO–NiO–TiO_2, *Thermochim. Acta*, **2**, 227–92 (1971).

3

Preparation and Characterization of Ceramic Powders

H. Schubert and G. Petzow

*Max-Planck-Institut für Metallforschung Institut für
Werkstoffwissenschaft, Stuttgart, FRG*

ABSTRACT

*Ceramic manufacturing based on powders, shape compaction and
subsequent densification is a process combining a number of different
independent steps. However, there is an extended number of examples
showing that the properties of the starting powder have a strong impact
on the final products' properties. Despite the high number of processing
steps the starting material used is still important; the result of using a
poor starting material cannot be rectified during further preparation.*

*$ZrO_2-Y_2O_3$ will be used as an example. Scanning transmission and
transmission electron micrography point analysis was used to characterize the chemical homogeneity, and in addition the properties of
importance for technical use are reported. The results of the characterization will be presented in the form of star-shaped diagrams. Each
property is plotted on an axis (non-satisfying properties inside, ideal
behaviour outside) and nine different properties are chosen.*

*It can be demonstrated that powders for technical use are charac-
terized by a single plot showing a well-balanced set of properties, but
none of the single properties is ideal. Until now sufficient materials for
compromise have been available but there is still large scope for
development.*

1. INTRODUCTION

Ceramic manufacturing is a process based on the compaction of
powder particles by pressing, casting, injecting, extruding or similar

techniques to form a compact which will subsequently be densified by a sintering treatment. This process involves in practice 10–15 independent manufacturing steps, often carried out by different experimentalists in different companies. The final quality of the product is of course determined by the preparation quality of every single step. But it has to be noted that the characteristics of the starting powder decide the characteristics of the final products. The expected microstructural properties of the product, such as chemical homogeneity, grain size, defect populations, and the presence or absence of an intergranular glassy phase, which are closely related with the physical and chemical properties (strength, toughness, corrosion resistance, thermal and electrical conductivity, etc.) are ultimately a consequence of the characteristics of the starting powder.

A couple of examples are given to illustrate the importance of powder characteristics and their impact on the final properties.

1.1. Oxygen in Si_3N_4

Si_3N_4 ceramics are used for several structural applications (cutting tools, motor parts) because of their relatively high strength even at elevated temperatures.

The preparation of Si_3N_4 requires the addition of oxidic sinter aid which forms a liquid phase. After cooling, the liquid is left in the microstructure as a glassy phase.

During service at elevated temperature, the glass phase will undergo the glass transformation, which promotes creep rupture [1, 2] at elevated temperatures (i.e. within the molten glassy phase a porosity evolves under stress, finally opening a crack path). The volume fraction of the glassy phase should therefore be kept as small as possible. One of the contributions of the glassy phase is in the surface oxygen content of the starting powders. Commercial Si_3N_4 is covered by an O_2 layer which—depending on the preparation route and individual batch—is equivalent to 0–2·5 wt% O_2 on the surface and up to 0·8 wt% O_2 in the lattice. Provided that this O_2 content is bound to Si cations, the Si_3N_4 powder is contaminated with up to 4 vol% of SiO_2. Without the presence of any strong reducing agent such as carbon, this SiO_2 content will stay in the material throughout the whole manufacturing process. Oxygen content therefore plays a dominant role in the formation of a glassy phase.

1.2. Chemical Homogeneity of ZrO$_2$–Y$_2$O$_3$ Ceramics (Y–TZP)

Y–TZP is a fine-grained ($<0.5\ \mu$m) single-phase ceramic. The effect of transformation toughening combined with the fine grain size and a defect-poor microstructure gives this material outstanding strength, toughness and damage tolerance ($\sigma > 1000$ MPa, $K_{Ic} = 10$ MPa$\sqrt{m}$ [3, 4, 5]). However, in the temperature range around 250°C in the presence of atmosphere containing H$_2$O vapor a degradation of the material, starting from the surface, takes place [6, 7, 8]. It can be shown that this degradation is accompanied by the transformation of the tetragonal grains to the stable monoclinic structure; hence, both grain size and grain size distribution, and also the stabilizer content and the distribution of the stabilizer, were identified as important microstructural parameters [9, 10]. A long time stability can be achieved when the grains are small and when the chemical composition is within a very narrow interval and higher than an oriented composition [10].

Both requirements can be satisfied when the starting powder is already homogeneous and exhibits a fine particle size. Hence, owing to the low sinter temperature and the short times, an inhomogeneity within the powder cannot be sufficiently overcome during densification. Again, the properties of the final component are predetermined by the powder characteristic.

2. CHARACTERIZATION AND ASSESSMENT OF ZrO$_2$–Y$_2$O$_3$ POWDERS

ZrO$_2$ powders will be taken as an example to discuss the differences in powder characteristic for different production routes and manufacturers.

The various preparation routes will briefly be described in the following paragraphs.

(1) The above alkoxide hydrolysis starts from an alcoholic solution in which the cations are perfectly distributed. During the hydrolysis, the alkoxides react to hydroxides and form a polymeric gel; in a subsequent calcination the oxides are formed.

The alkoxides can be purified by distillation, which enables the preparation of extraordinarily cation-pure precursors. However, for large-scale manufacturing, inert-atmosphere processing, which is re-

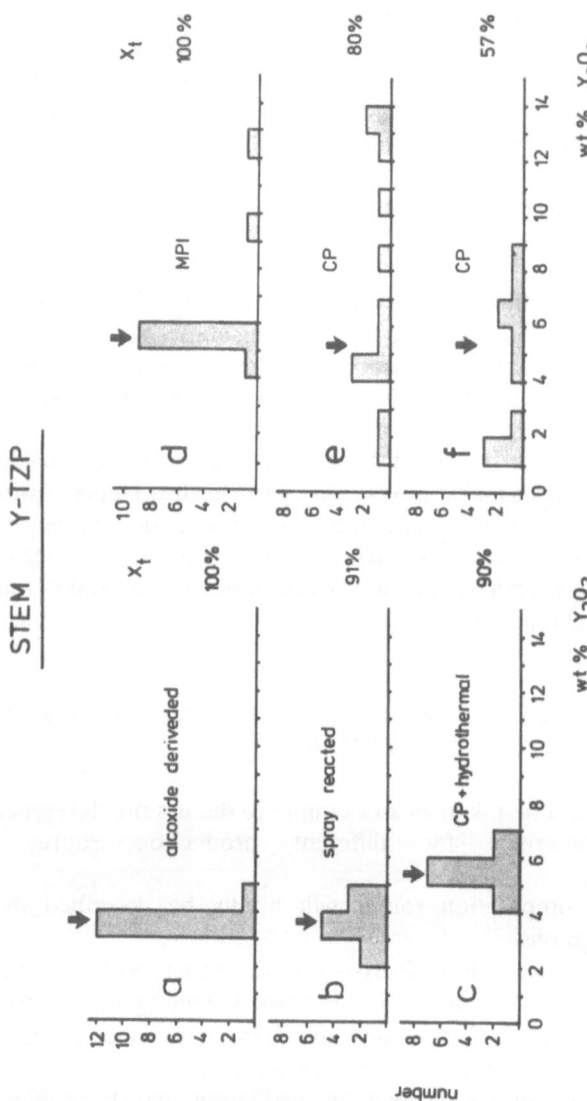

FIG. 1. Y_2O_3 distribution in Y-TZP powders. The number of measurements within an interval of 1 wt% Y_2O_3 is plotted versus the Y_2O_3 content. x_t = tetragonal content by X-ray diffraction.

quired for the oxidation sensitive alkoxides, is not preferred. Hence, powder producers have been reluctant to use this process because of the high price of alkoxides.

The powders prepared from the alkoxide precursors are known to be chemically very homogeneous. Even for a calcination temperature of only 500°C [11] the material was found to be single-phase tetragonal and scanning transmission electron microscopy (STEM) measurements showed that the Y_2O_3 content of the powder particles varied only 1 wt% either side of the theoretical composition weight (Fig. 1). The crystallite size was approximately 5 nm (Fig. 2). However, this extremely fine particle size caused a very low taping density and, as a consequence, high shrinkage both during pressing and during sintering occurred. Without further powder processing it would be very difficult to handle such a powder in commercial manufacturing.

(2) Aqueous salt solutions are decomposed directly to oxide in *reaction spray drying*. A nozzle is used to produce small droplets which are blown into a ~1000°C combustion furnace [12]. The outer shell of the droplets is heated in a few tenths of a second and the oxide crystallizes in the shape of the droplet. Finally, the interior liquid is heated and evaporated and bursts the shell. The agglomerate of reaction spray-dried powders typically has hollow-sphere geometry.

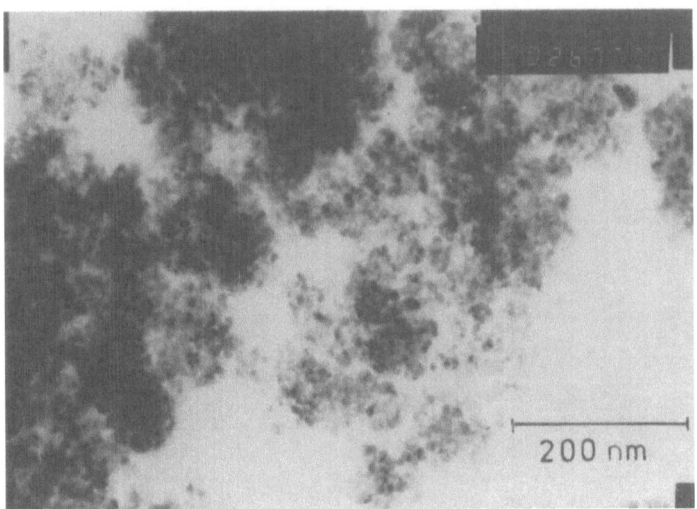

FIG. 2. TEM bright field. Alkoxide hydrolysis ((a) in Fig. 1).

Owing to the high speed of crystallization, these powders are chemically homogeneous, i.e. the compositions measured by STEM varied only within 2·5 wt% (Fig. 1).

The crystallite size was approximately 70–100 nm and the crystallites were densely aggregated (Fig. 3). It required 5 hours in an attritor to break up these aggregates.

(3) *Hydrothermal crystallization* is a variant of precipitation [13]. Salt solutions are either precipitated under hydrothermal conditions or are separately precipitated to hydroxides and subsequently crystallized under hydrothermal conditions. For this investigation the latter method was applied, which led to a powder whose chemical homogeneity is comparable to that of the spray-reacted powder (Fig. 1). The crystallite size was ~100 nm (Fig. 4) and the crystallites were loosely agglomerated to 10–50 μm flakes. ·

(4) The commercially most advanced method is *co-precipitation.* Solutions are precipitated by ammonia, urea or oxalic acid to form hydroxides or oxalates which are subsequently converted to oxides by calcination. The homogeneity of the powder depends on the method of precipitation, because Zr and Y salts precipitate at different pH values, which may cause a separation even during the first precipitation step. The two tested commercial powders were chemically not

Fig. 3. TEM bright field. Densely sintered crystallites of a spray reacted material ((b) in Fig. 1).

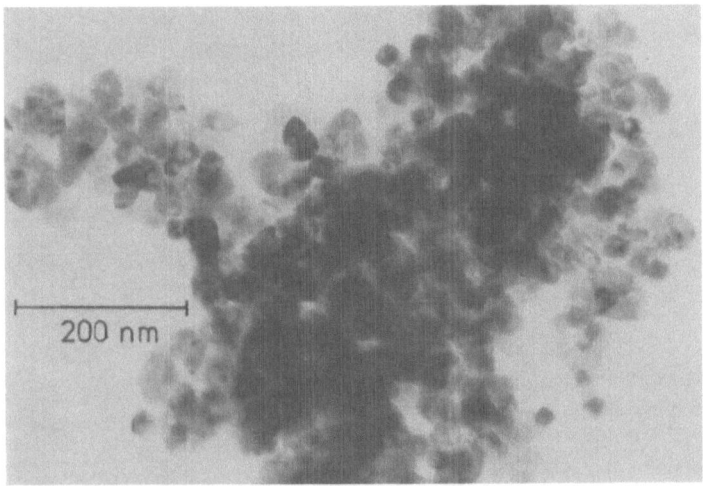

FIG. 4. TEM bright field. Hydrothermal crystallization powders have fairly separated crystallites ((c) in Fig. 1).

very homogeneous (tetragonal content by X-ray diffraction $x_t = 57\%$ and $x_t = 80\%$) and for both wide variations in the chemical composition were measured by STEM (Fig. 1, Fig. 5).

The crystallites of ~50–100 nm size were connected to flaky agglomerates (100 μm). These powders could be packed to green bodies of only 40–50% TD (theoretical density), not sufficient for a commercial process. Co-precipitated powders are usually milled and spray-dried before further processing.

The chemical homogeneity of co-precipitated powders can be improved by faster precipitation (powder MPI in Fig. 1). The material is then obtained as single-phase tetragonal even for calcination temperatures as low as 500°C ($x_t = 100\%$) and the variation of the chemical composition is almost as small as in the alkoxide-derived material. The crystallite size is only 5 nm (Fig. 6) and, in consequence, the green density is low, but both pressing shrinkage and sintering shrinkage are high. A higher calcination temperature and a subsequent milling treatment are required in order to obtain sufficient pressability.

Most research activities have been focused both on fine crystallite size and chemical homogeneity and, hence, they have opened new routes for powder processing. However, for the actual preparation of TZP components there are more parameters requiring attention than

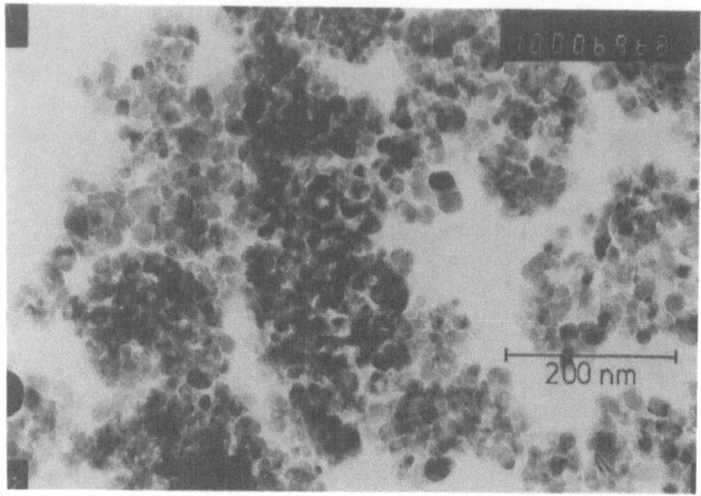

FIG. 5. TEM bright field. Co-precipitated powders ((e) in Fig. 1).

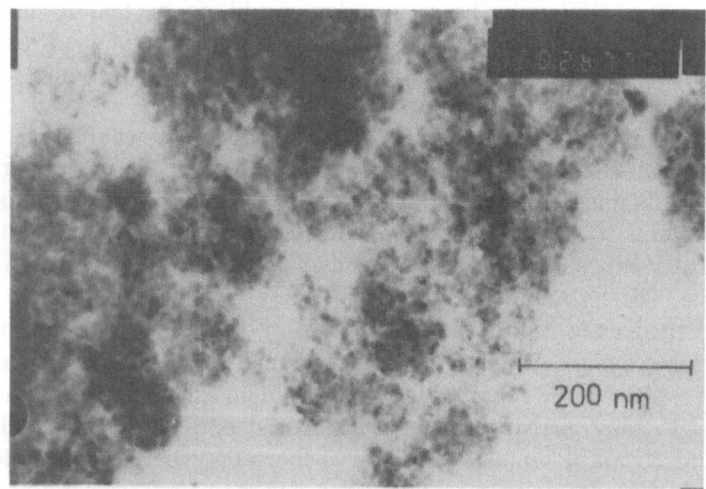

FIG. 6. TEM bright field. Co-precipitated powders ((d) in Fig. 1).

just crystallite size and homogeneity. A large shrinkage on sintering is to be avoided because it would be very difficult to produce crack-free large components. The high sinter activity that is prefered because of the low sinter temperature is a result of the small crystallite size, but the high sintering shrinkage is also a consequence of these small crystallites. Finally, there are other contradictory powder characteristics which all have to be addressed.

The various powder characteristics of zirconia will be displayed in terms of a star diagram showing (a) chemical homogeneity, (b) crystallite size, (c) crystallite shape, (d) particle size, (e) particle-size distribution, (f) green density, (g) pressing shrinkage, and finally (h) sintering shrinkage. The scale on these diagrams has been chosen to show the differences between various powders, however, it can be modified for any other type of powder and application (Si_3N_4, SiC,

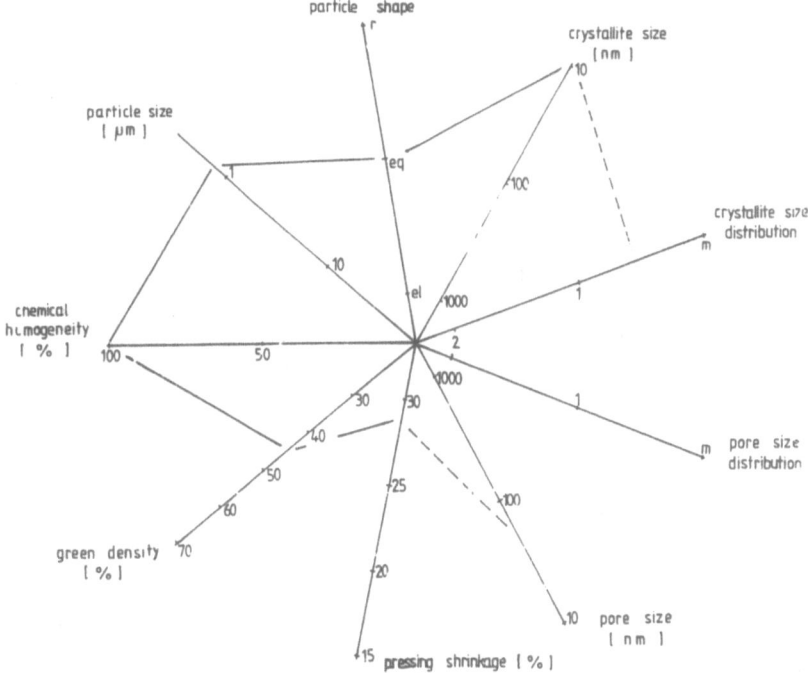

FIG. 7a. Star diagram of different powder properties. Alkoxide powders are chemically homogeneous and have very fine crystallites but their technical handleability is difficult.

etc.). The properties of an 'ideal' powder, which would satisfy the requirement of homogeneity and the requirements of a technical manufacturing process, would be represented by an outer circle on the star diagram, and a 'poor' powder would be plotted on the centre point.

A mixed oxide powder contains two unreacted phases. The chemical homogeneity cannot be expressed by a Y_2O_3-distribution curve. The homogeneity of the sintered body depends mostly on the diffusion during sintering.

For many applications this would not be sufficient (corrosion resistance of TZP [10]). The particle size is sufficiently small and the achieved green density is comparably high owing to the removal of aggregates. Hence, both pressing shrinkage and sintering shrinkage are low (Fig. 7a).

In contrast to mixed oxide powder, the alkoxide-derived materials

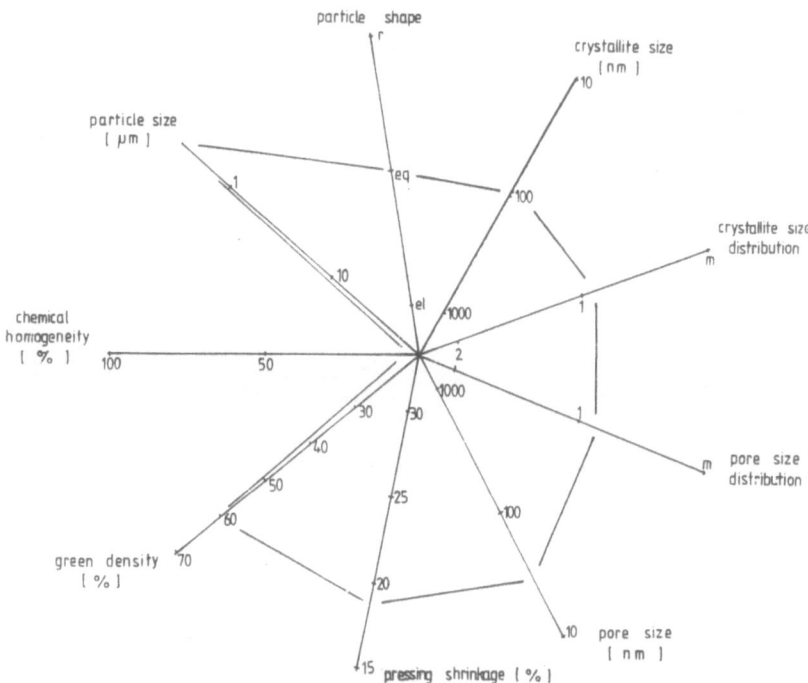

FIG. 7b. Star diagram of different powder properties. Oxide mixtures are chemically completely inhomogeneous but they can be handled easily.

are perfectly homogeneous (compare Fig. 1) but their technical usability is limited by their extraordinarily low green densities and the high shrinkages (Fig. 7b). The 'engineering' properties of these powders makes them too difficult to handle, and as a consequence they have not yet found a large-scale market.

The materials which are nowadays commonly distributed on the ceramic market are co-precipitated materials. Their chemical homogeneity is not as high as the alkoxide-derived material but is high enough to give a homogeneous body after sintering. The morphological properties are not as good as simple attrition-mixed oxide but they are good enough to make the material useful for technical processes. The plot of the properties in the star diagrams (Fig. 7c) is almost a perfect circle but there are still possibilities for improving each parameter.

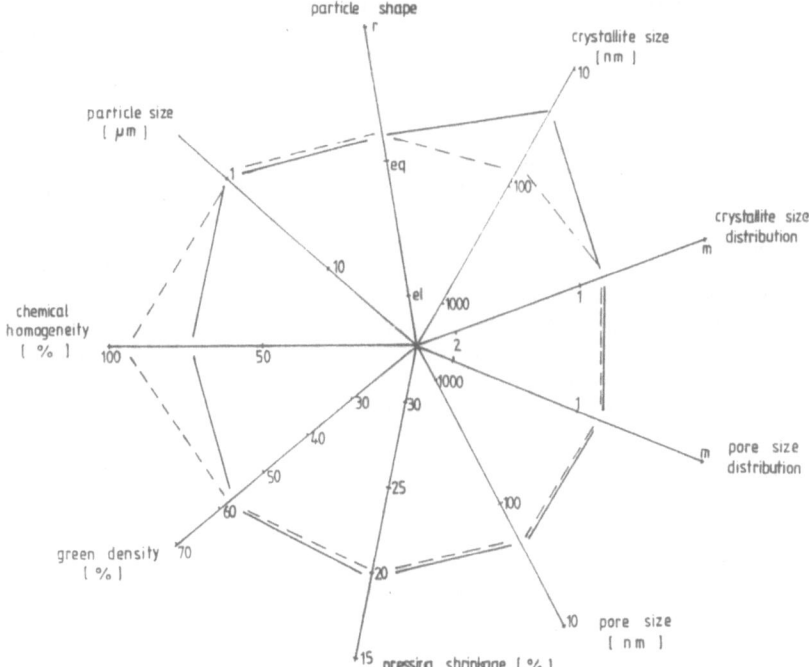

FIG. 7c. Star diagram of different powder properties for co-precipitated (solid line) and hydrothermally crystallized powder (broken line) are characterized by a well-balanced set of properties, but none of the properties is ideal. They are the favorable compromise at present.

These materials are engineered to a satisfying compromise between homogeneity particle size on one side and the technical properties (shrinkages, densities) on the other.

Finally, it can be stated that materials with balanced properties are available but that there are possibilities for further improvement in all properties. Powder development has still a great future.

REFERENCES

1. P. Greil, Ph.D. Thesis, University of Stuttgart, 1982.
2. M. Matsui, Y. Ishida, T. Soma and I. Oda, Ceramic Turbocharges Rotor Design Considering Long Term Durability. In *Ceramic Materials and Components for Engines,* ed. W. Bunk and H. Hausner, Verlag Deutsch Keramische Gesellschaft, pp. 1034–42.
3. M. Matsui, T. Soma and I. Oda, Effect of microstructure on the strength of Y-TZP components. In *Advances in Ceramics 12,* ed. N. Claussen, M. Rühle and A. H. Heuer. The American Ceramic Society, Columbus, Ohio, 1984, pp. 371–81.
4. K. Tsukuma, Y. Kubota and T. Tsukidate, Thermal and mechanical properties of Y_2O_3-stabilized tetragonal zirconia polycrystals. In *Advances in Ceramics 12,* ed. N. Claussen, M. Rühle and A. H. Heuer. The American Ceramic Society, Columbus, Ohio, 1984, pp. 382–90.
5. M. Matsui, M. Masuda, T. Soma and I. Oda, Thermal stability and microstructure of Y-TZP. In *Advances in Ceramic 24,* ed. S. Sōmiya. The American Ceramic Society, Columbus, Ohio, 1988, pp. 607–14.
6. W. Watanabe, S. Iio and I. Fukuura, Aging behavior of Y-TZP. In *Advances in Ceramics 12,* ed. N. Claussen, M. Rühle and A. H. Heuer. The American Ceramic Society, Columbus, Ohio, 1984, pp. 391–8.
7. M. Yoshimura, T. Noma, K. Kawabata and S. Somiya, Role of H_2O in the degradation process of Y-TZP, *J. Mater. Sci. Lett.,* **6,** 465–76 (1987).
8. S. Schmauder and H. Schubert, Significance of internal stresses for the martensitic transformation in yttria-stabilized tetragonal zirconia polycrystals during degradation, *J. Am. Ceram. Soc.,* **69**(7), 534–40 (1986).
9. H. Schubert, Investigations on the stability of yttria stabilized tetragonal zirconia (Y-TZP), *Zirconia Ceramics* **7,** 65–81 (1985).
10. H. Schubert and G. Petzow, Microstructural investigations on the stability of yttria stabilized tetragonal zirconia, *Advances in Ceramics 24,* ed. S. Sōmiya, The American Ceramic Society, Columbus, Ohio, 1988, pp. 21–8.
11. H. Schubert and G. Petzow, Preparation and characterization of multi-component ceramic powders. In *Science of Ceramics 14,* ed. D. Taylor. The Institute of Ceramics, Stoke-on-Trent, UK, 1988.
12. R. Schmittberger, Pulvertechnologie auf neuen Wegen, Dornier, Friedrichshafen, FRG.
13. E. Tani, M. Yoshimura and S. Sōmiya, Formation of ultrafine tetragonal ZrO_2 powder under hydrothermal conditions, *J. Am. Ceram. Soc.,* **66**(1), 11–14 (1983).

4

Contributions of Sintering and Coarsening to Densification: A Thermodynamic Approach

F. F. LANGE

*Materials Department, College of Engineering, University of
California, Santa Barbara, California, USA*

ABSTRACT

Past sintering concepts detailed the mass transport for two particles, but because two particles cannot define a pore, they ignored the void phase within the powder compact defined by the network of touching particles. Recent thermodynamic treatments of particle networks have shown that densification is not only limited by mass-transport kinetics, but is also limited by thermodynamic considerations. Specifically, these treatments show that the network can only disappear as the configuration changes through coarsening phenomena. That is, coarsening phenomena reinitiate sintering and lead to densification. These new concepts predict that densification is predictably related to grain growth.

1. INTRODUCTION

'When we know enough about sintering we will greatly enhance our understanding of all of nature.' Rhines [1] statement most concerns ceramists, who, by-and-large make engineering components from powder compacts made dense by a heat treatment.

Densification is driven by the excess free energy associated with a powder's surface area. The greater the surface area of the powder, i.e. the smaller the particle size, the greater the thermodynamic potential for densification. This potential is only realized when mass is free to roam and relocate. Mass relocation is limited by the kinetics of atomic

57

and molecular species, which is greatly enhanced at elevated temperatures.

Paths of mass relocation are dictated by different surface curvatures existing within the powder compact as described by the Gibbs–Kelvin equation,

$$\Delta\mu = \Omega\gamma_s\left(\frac{1}{R_2} - \frac{1}{R_1}\right) \tag{1}$$

which relates the chemical potential difference ($\Delta\mu$) between two surface locations (1 and 2) to their respective surface curvatures, expressed by their combined orthogonal principal radii of curvature R_1 and R_2, the surface energy per unit area (γ_s), and the atomic volume (Ω) of the mobile species. When one defines surface 2 at or near the contact perimeter between two touching particles and surface 1 as any other position on the particle surface, R_2 is negative and small and R_1 is positive and large. Because $|1/R_2| \gg -|1/R_1|$, the free energy of the system will decrease ($\Delta\mu$ is negative) when mass relocates to the region where the two particles make contact. This mass relocation causes a neck to grow between every pair of touching particles. This process, known as sintering, was first demonstrated by Kuczynski [2], who in 1949, sintered large polycrystalline particles onto flat polycrystalline substrates.

As reviewed by Exner [3], theories developed to detail sintering are based on two touching spheres. They were developed to determine the rate of neck growth and, for appropriate mass transport paths, the rate that particle centers approach one another to cause shrinkage. These theories assume that the flux of mass to the neck is proportional to $\Delta\mu$. Using approximate geometric solutions for the radius of curvature at the neck ($|R_2|$ proportional to $|R_1|$), these theories conclude that the rate of neck growth is inversely proportional to particle size raised to a power that depends on the mass transport mechanics, viz., the rate of neck growth is proportional to $(1/R_1)^m$. That is, the smaller the particles, the faster the sintering kinetics at any given temperature.

Many different mass transport mechanisms were detailed for crystalline particles: volume diffusion, surface diffusion, grain-boundary diffusion, evaporation-condensation, and liquid solution/reprecipitation. It was concluded, *a priori*, that only mechanisms that relocate mass from between particle centers to the neck, viz., grain-boundary and/or volume diffusion, would allow particles to shrink together. Shrinkage implies densification. Likewise, it was assumed that mechanisms which relocate mass from the particle's surface to the neck, via evaporation–condensation and/or surface diffusion, could not produce shrinkage and densification. As shown below, this conclusion is not true if the exterior to interior surface path of a powder compact is short and open.

Sintering theories based on neck growth have never successfully described the complete densification history of a powder compact [3]. To overcome this deficiency, separate theories have been developed to explain different stages of densification. Neck growth is commonly classified as the initial stage of densification. After necks grow between the touching particles, the observed morphology of powder compacts continuously changes from one of touching particles and interconnected porosity to one of closed, isolated pores. During this intermediate stage, particles, which are now called grains, are observed to coarsen (some disappear, others grow). Current theories do not incorporate this coarsening phenomenon. As discussed below, coarsening will dominate mass transport kinetics after neck growth between the initial, touching particles terminates. In addition, coarsening leads to morphological changes in the partially dense network such that neck growth is reinitiated within the changing network and densification proceeds with coarsening.

Many theories have also been developed for the final, closed-pore stage of densification. These theories predict that pores, assumed to have concave surface curvatures (looking from within) and located at 2, 3, or 4 grain junctions, continuously shrink at a rate controlled by the kinetics of either volume and/or grain-boundary diffusion. Microstructural observations show that many of the isolated pores have convex curvatures and, as discussed below, can neither shrink nor grow without grain growth.

These and other new contributions to densification theory result when one includes the free-energy change of grain boundaries with that of the particle surface for particle configurations that encompass void phase.

2. EQUILIBRIUM CONFIGURATION OF SINTERED PARTICLE ARRAYS

Two touching particles are certainly useful in modeling neck growth, but since they do not enclose space, i.e. do not define a pore, they cannot be used to model how the void phase disappears within a powder compact. As pointed out by Kellett and Lange [4], the conditions for void disappearance must be modeled with particle arrays that enclose porosity. The simplest of these arrays is a linear array of identical spheres joined end-to-end to form a ring. Although the ring array is emphasized below, the same conclusions result for three-dimensional arrays, e.g. touching spheres that form regular polyhedra.

The free energy of simple arrays of identical crystalline particles was determined as a function of their configurational change during mass transport to form necks. The energy per particle was determined by summing the energies associated with its surface area (A_s) and its grain boundary area (A_b) formed with adjacent particles. This is expressed as

$$E = \gamma_s A_s + \gamma_b A_b \tag{2}$$

where γ_s and γ_b are the energies per unit area for the surface and the grain boundary, respectively. It was assumed that γ_s and γ_b are related through Young's equation

$$\gamma_b = 2\gamma_s \cos\left(\frac{\varphi_e}{2}\right) \tag{3}$$

where φ_e is the equilibrium (dihedral) angle formed where the grain boundary intersects the particle's surface. Equation (1) can be rewritten as

$$E = \gamma_s\left(A_s + 2\cos\left(\frac{\varphi_e}{2}\right) A_b\right) \tag{4}$$

By differentiating eqn (3), it can be seen that the grain boundary will only grow ($dA_b > 0$) when the surface area decreases ($dA_s < 0$), and it will cease to grow when

$$\frac{dA_s}{dA_b} = -\frac{\gamma_b}{\gamma_s} \tag{5}$$

i.e. when $dE = 0$. The configuration of the array when the contact area ceases to grow is defined as the equilibrium configuration.

As defined, the development of the equilibrium configuration precludes mass transport between particles, i.e. particles should maintain their initial mass during sintering. To determine the energy of the array as the contact area grows, one needs to define an initial particle shape, how this changes, and the relation between the particle's surface area and its grain-boundary area during mass transport. Since mass transport is a kinetic phenomenon, in reality, the particle's surface curvature will be a function of position and time. Such a non-uniform surface curvature would make the definition of the particle shape analytically intractable. Therefore, to allow analytical calculations, it is assumed that the surface curvature is independent of position as the array develops its equilibrium configuration, i.e. the initial touching spheres remain spherical as they interpenetrate, but increase their radius to conserve mass. Relative to the real system (surface curvature is some function of position), this assumption will result in a different array energy (expected to be lower) as mass transport leads to equilibrium. But, although the assumed particle configurational change used to calculate the array's energy will be different from that of the real system during mass transport to achieve equilibrium, both will be identical at equilibrium.

Several geometric variables can be used to describe configurational change of an array during neck growth. They include the angle defined by the tangents to the external surface at the grain boundary, termed the contact angle (φ, see Fig. 1); the distance between mass centers;

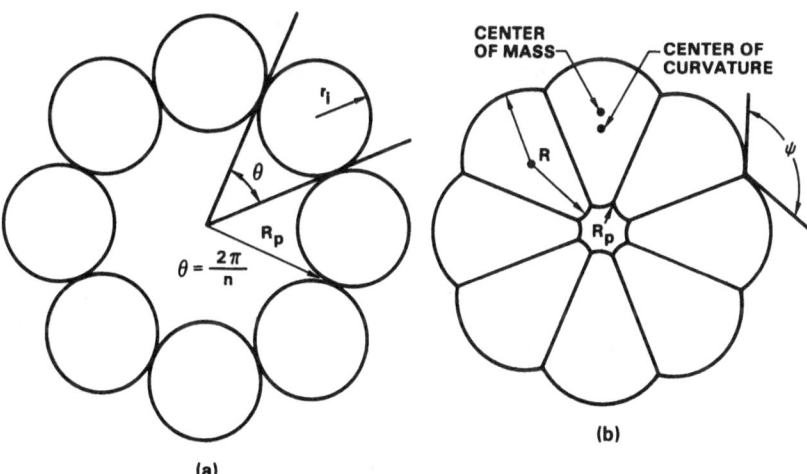

FIG. 1. Schematic of a ring of identical, spherical, touching particles (a) that develop an equilibrium configuration (b) by neck growth.

and the grain boundary radius. For arrays that define a single pore, other geometrical variables are needed both to define the configuration and to calculate array energies. These variables include the radius (R_p) of the pore enclosed by the particles, and the number (n) of particles coordinating the pore. The pore radius can be be defined by a circle (or sphere) that circumscribes the positions where the grain boundaries intersect the pore's surface, as shown in Fig. 1. The pore's coordination number n is an independent variable and can be described by the angle that subtends a single particle. For ring arrays, the coordination angle is simply related to the coordination number, $\theta = 2\pi/n$.

The radius of curvature (r) of the particle surface is uniquely defined by its initial radius (r_i) and the contact angle (φ) for the case where the pore within the ring does not close $(R_p > 0)$:

$$r = r_i \left[\frac{1}{2} \cos \frac{\varphi}{2} \left(2 + \sin^2 \frac{\varphi}{2} \right) \right]^{-1/3} \tag{6}$$

For the same conditions $(R_p > 0)$, the energy per particle can be expressed as

$$E = 4\pi r_i^2 \gamma_s \frac{\cos \frac{\varphi}{2} + \frac{1}{2} \sin^2 \frac{\varphi}{2} \cos \frac{\varphi_e}{2}}{\left[\frac{1}{2} \cos \frac{\varphi}{2} \left(2 + \sin^2 \frac{\varphi}{2} \right) \right]^{2/3}} \tag{7}$$

and the radius of the pore can be expressed as

$$R_p = \frac{\cos \left(\frac{\Psi}{2} + \frac{\pi}{n} \right)}{\sin \left(\frac{\pi}{n} \right)} \left[\frac{1}{2} \cos \frac{\varphi}{2} \left(2 + \sin^2 \frac{\varphi}{2} \right) \right]^{-1/3} r_i \tag{8}$$

Other relations for r and E are found when $R_p < 0$ [4], but they are not important for the following discussion.

A plot of the particle energy, normalized by its initial energy, as a function of the pore size, R_p, is illustrated in Fig. 2 for the case where $\varphi_e = 120°$ $(n_c = 6)$ for ring arrays containing different numbers of particles. As shown, when the number of particles, n, is larger than a critical number, n_c, the pore within the array shrinks to an equilibrium size defined by a minimum in the array's energy function. When $n \le n_c$, the pore shrinks and disappears. By examining eqn (8), it can

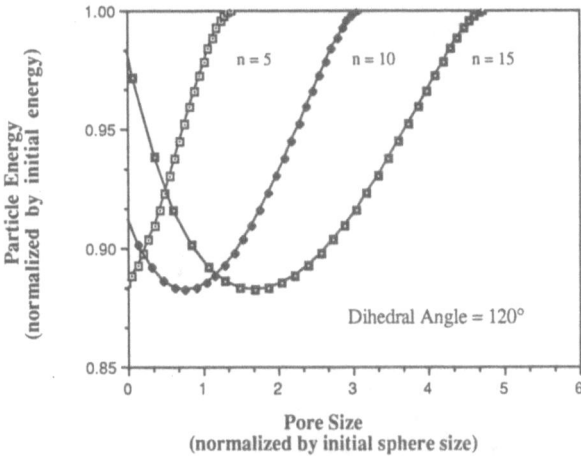

FIG. 2. Normalized energy of the ring configuration as a function of pores size for the case where 5, 10 and 15 particles coordinate the ring and a dihedral angle of 120°.

be seen that $n_c = 2\pi/(\pi - \varphi_e)$. These calculations show that if mass transport does not take place between particles (coarsening does not occur), a pore enclosed by sintering particles will either shrink to an equilibrium size (case where $n > n_c$) or disappear (case where $n \leq n_c$). These calculations strongly suggest that not all pores within a powder compact will disappear during sintering. That is, some pores (i.e. those with $n > n_c$) will only shrink to an equilibrium size.

It should be noted that no specific mass transport mechanism or path was defined to obtain the above results. Thus, if a connective path remains open from the exterior surface of the array to the pore surface, surface diffusion and/or evaporation–condensation mechanisms can cause array and pore shrinkage. This result cannot be achieved for the linear array, showing one distinct difference between linear arrays and arrays that enclose a pore.

3. COARSENING DURING DENSIFICATION

As illustrated above, a particle array will sinter to an equilibrium configuration when eqn (5) is satisfied. It can be shown that when eqn (5) is satisfied, the curvature at the neck is identical to the particle's

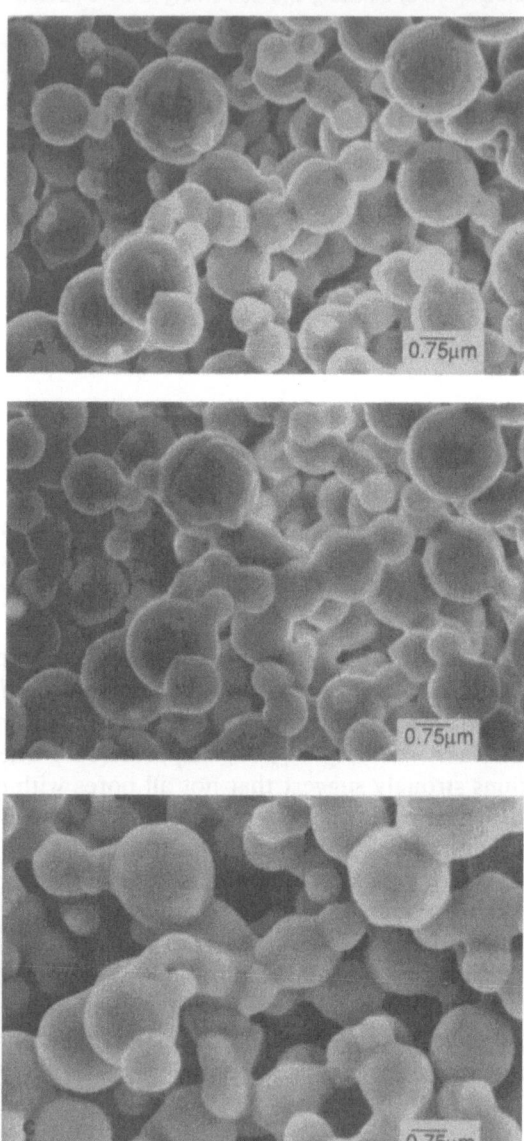

FIG. 3. Micrographs of spherical ZrO_2 (+8 mole% Y_2O_3) particles produced by electrostatic atomization of Zr-acetate. (a) After sintering by heating to 1300°C for 10 h. (b) Same area after further heat treatment at 1300°C for 18 h and (c) 1400°C for 4 h where a self-similar network produced by initial sintering undergoes coarsening and further densification. Note the change in two smaller particles that join larger particles in upper left hand corner, which leads to non-equilibrium structure after coarsening.

surface curvature (i.e., $R_2 = R_1$), and thus, the potential for sintering (i.e. neck growth), dissipates to zero, that is, $\Delta\mu_{sintering} = 0$. The above discussion also shows that some pores $(n > n_c)$ may not disappear during this sintering process. Yet an experimentalist knows that nearly all pores can be made to disappear by choosing a proper temperature–time schedule. How then is this experimental observation consistent with the above theoretical results? The clue to answer this question lies with microstructural observations.

The sintered network of touching ZrO_2 spherical particles developed by heating at 1300°C for 10 h is shown in Fig. 3(a). Note in Fig. 3(b) and 3(c) that when the same area is viewed after a subsequent heat treatment at 1300°C for 18 h and 1400°C for 4 h, respectively, the particle network appears self-similar. Differences between the micrographs can be seen. First, smaller particles (or grains) either become smaller or disappear, while larger particles become larger, i.e. coarsening has occurred. Second, groups of grains are rearranged relative to others. Third, measurements show that some shrinkage, i.e. densification, has occurred. And fourth, very few new contacts are made. The more comprehensive study [5] from which these micrographs are taken show that once the initial touching particles sinter together to form a metastable network, further densification is related to grain coarsening, which continuously alters the network configuration.

Although mass transport to the contact region may stop when eqn (5) is satisfied, adjacent, sintered grains will have different radii of curvature that will drive interparticle mass transport, as described by eqn (1), where R_1 and R_2 are now the surface curvatures of adjacent particles. Interparticle mass transport will cause coarsening, i.e. smaller grains disappear as larger grains grow. Since the radii of both particles are positive and much larger than the negative radius where particles touch, the driving potential $(\Delta\mu_{coarsening})$ and related mass flux for interparticle mass transport are much smaller than those associated with transport to the neck region. Thus, one might expect that the kinetics for interparticle mass transport leading to coarsening (grain growth) will be much slower than the transport to the contact regions (sintering). That is, although the neck growth and coarsening phenomena are concurrent, sintering will dominate until it nearly terminates by satisfying eqn (5).

The question of how coarsening is related to the thermodynamics of pore stability can be viewed in two different but complementary ways.

First, it can be seen that coarsening will decrease the number of grains coordinating stable pores, i.e. convert a stable pore with $n > n_c$ to an unstable pore with $n \leq n_c$. With this view, Kellett and Lange [6] have shown that isolated pores produced in a powder compact when the polymer spheres pyrolyze during heating will only disappear when grain growth decreases their coordination number below a critical value.

Second, the relation between coarsening, configurational changes in the network during coarsening, and shrinkage can be obtained by examining what happens when grains disappear within a sintered network. Figure 4(a) illustrates three truncated, spherical particles taken from a network which has shrunk to a metastable configuration by sintering. The dihedral angle φ_e defines the equilibrium between the surface and grain boundary energies which will be achieved when eqn (5) is satisfied, viz., $2\cos(\varphi_e/2) = \gamma_b/\gamma_s$. As coarsening proceeds, the smaller grain becomes smaller as neighboring grains become larger. At some point during coarsening, neighboring grains touch one another, as illustrated in Fig. 4(c). It can be shown [4] that when the neighboring grains touch, the angle formed with the surface tangents and the new grain boundary is less than φ_e. Since $\varphi < \varphi_e$, eqn (5) is no

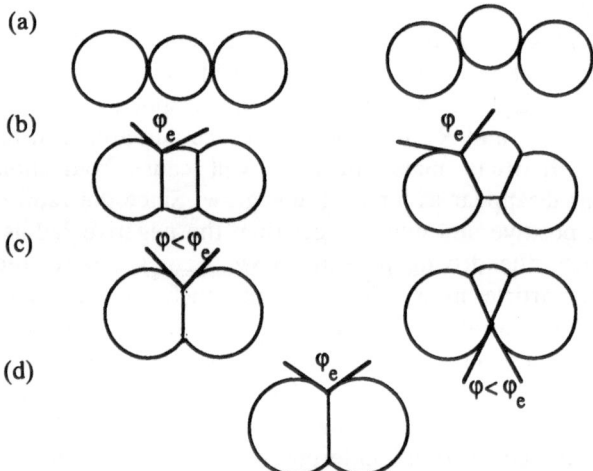

FIG. 4. Three particles, in two configurations, which (a to b) shrink together by sintering and (b to c) coarsen (center particle disappears) and then (c to d) sinter again.

longer satisfied and mass transport to the contact region, i.e. sintering, will reinitiate. Contrary to intuition, it can also be shown [7] that very little shrinkage occurs as the smaller particle disappears. That is, for pertinent values of γ_b/γ_s, the larger particles grow at the same rate as that at which the smaller one shrinks. Once the larger particles touch one another, shrinkage occurs by the reinitiation of sintering until a new metastable configuration develops that satisfies eqn (5) as shown in Fig. 4(d). Thus, although most of the mass transport period is consumed by coarsening, coarsening reinitiates sintering and shrinkage. At the same time, coarsening decreases the pore coordination number.

If we argue that rapid transport to the contact region leads to shrinkage by sintering and the development of a metastable network similar to that shown in Fig. 4(a), and that further shrinkage is controlled by slower interparticle transport, then the kinetics of densification should be separated by two regimes: an initial regime controlled by sintering kinetics and a subsequent regime controlled by coarsening kinetics. Since the sintering regime results in shrinkage to a metastable network with a given relative density, further densification must be controlled by coarsening kinetics. These two regimes will be separated by the relative density of the metastable network produced by sintering.

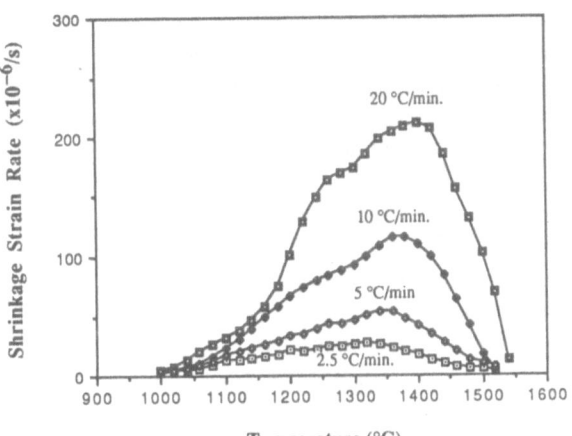

FIG. 5. Shrinkage strain rate vs temperature for identical Al_2O_3 compacts heated to 1550°C at different heating rates. Maximum strain rate occurred at a relative density of 0·77 for all specimens.

Figure 5 shows the shrinkage strain rate determined when different specimens cut from a single powder compact of Al_2O_3 were heated to 1550°C at different heating rates [8]. In each case, the shrinkage strain rate increases to a maximum and then decreases. The maximum shrinkage strain rate corresponds to the inflection in relative density vs temperature curves observed in densification experiments. For each curve, the maximum shrinkage strain rate occurs at the same relative density of 0·77. These data strongly suggest that sintering kinetics dominate up to a relative density of 0·77, where coarsening kinetics dominate during further densification.

4. STABILITY AND INSTABILITY CONDITIONS FOR ISOLATED PORES

Kingery and Francois [9] were the first to use the fact that isolated pores intersected by grain boundaries within a polycrystalline material have surface curvatures that depend on the number of coordinating grains, n, and the dihedral angle. Recognizing this fact, they suggested that mass transfer either into or away from a pore will depend on its surface curvature. They concluded that if the number of grains coordinating the pore is less than a critical value ($n < n_c$), then the pore's surface will be concave (looking from within the pore), which will promote mass transport to the pore's surface and pore closure. They also concluded that if the pore's coordination number is greater than a critical value ($n > n_c$), its surface will be convex, promoting mass transport away from the pore and thus, pore growth. The critical pore coordination number separating pore closure from pore growth is dependent on the dihedral angle. As it will be seen below, isolated pores with $n < n_c$ are unstable and will disappear as suggested by Kingery and Francois, but when $n > n_c$, the pores will either shrink or grow to a stable size where their surface curvature is identical to grains on the external surface.

As detailed by Kellett and Lange [7], energy calculations show that the change in free energy of an isolated pore with respect to its volume (V_p) is given by

$$\frac{\partial E_p}{\partial V_p} = \frac{2\gamma_s}{r_p} \tag{9}$$

where r_p is the radius of curvature of the pore surface. For pores with concave surfaces, r_p is negative and eqn (9) shows that the pore will decrease its energy by decreasing its volume to disappearance. This condition corresponds to pores with $n < n_c$. For pores with convex surfaces $(n > n_c)$, r_p is positive; eqn (9), by itself, suggests that the pore will grow. But, as material is either added or taken away from the external surface to either increase or decrease the internal pore, the free energy change of the surface and its intersecting grain boundaries must be taken into account. It can be shown that the change in free energy of the external surface as volume is either added or subtracted from the surface to change the pore volume is given by

$$\frac{\partial E_s}{\partial V_p} = -\frac{2\gamma_s}{r_s} \tag{10}$$

where r_s is the curvature of the surface grains, which is always convex, that is, r_s is always positive. Either by summing eqns (9) and (10) to obtain the free-energy change of the body (surface and pore), or by simply invoking eqn (1), it can be seen that isolated pores with concave curvatures (r_p negative, $n < n_c$) are always thermodynamically unstable and will disappear, kinetics permitting. On the other hand, isolated pores with convex curvatures (r_p positive, $n > n_c$) will either shrink or grow until $r_p = r_s$. It can be shown that the size of the stable, isolated pores will be related to the size of the surface grains [7]. Thus, isolated pores with $n > n_c$ will eventually achieve an equilibrium size unless they lower their coordination number through grain growth.

5. CONCLUDING REMARKS

The thermodynamic arguments presented above show that a greater understanding of densification can still be achieved after more than 40 years of intense study by the ceramics community. These arguments deal with the void phase, which is what we want to eliminate, as well as the particles. They teach that particle packing is crucial to densification phenomena, by showing that the ultimate objective is to form a powder compact where all pores are defined by $n < n_c$. If one cannot achieve this objective, they show that sintering alone (neck growth) will not produce a fully dense body, but coarsening and *controlled* grain growth are desirable and must be encouraged. It is hoped that this new thinking will not only aid our understanding of densification, but will also lead to innovative powder processing.

ACKNOWLEDGEMENT

The writing of this review was supported by the Air Force Office of Scientific Research under contract No. AFOSR-87-0291.

REFERENCES

1. F. N. Rhines, In *Science of Sintering and its Future*, ed. M. M. Ristic. International Institute for the Science of Sintering, Belgrade, 1975.
2. G. C. Kuczynski, 'Self diffusion in sintering of metallic particles, *Met. Trans. AIME* **185**, 169–78 (1949).
3. H. E. Exner, Principles of single phase sintering, *Reviews on Powder Metallurgy and Physical Ceramics* **1** (1–4), 1–251 (1979).
4. B. J. Kellett and F. F. Lange, Thermodynamics of densification, Part I: Sintering of simple particle arrays, equilibrium configurations, pore stability, and shrinkage. *J. Am. Ceram. Soc.*, **72**, 725–34 (1989).
5. E. B. Slamovich and F. F. Lange, Spherical zirconia particles via electrostatic atomization: Fabrication and sintering characteristics. *Mat. Res. Soc. Proc. Sym. on Ultrastr. Proc.*, **121**, 617–21 (1988).
6. B. J. Kellett and F. F. Lange, Thermodynamics of densification, Part III: Experimental relation between grain growth and pore closure, to be published.
7. F. F. Lange and B. J. Kellett, Thermodynamics of densification, Part II: Grain growth in porous compacts and relation to densification. *J. Am. Ceram. Soc.*, **72**, 735–41 (1989).
8. F. F. Lange, Powder processing science and technology for increased reliability. *J. Am. Ceram. Soc.*, **72**, (1) 3–15 (1989).
9. W. D. Kingery and B. Francois, Sintering of crystalline oxides, I. Interaction between grain boundaries and pores. In *Sintering and Related Phenomena*, ed. G. C. Kuczynski, N. A. Hooton, and G. F. Gibbon. Gordon & Breach, New York, 1967, pp. 471–98.

5

Ferroic Materials and Composites: Past, Present and Future

L. ERIC CROSS

*Materials Research Laboratory, The Pennsylvania State University
University Park, Pennsylvania, USA*

ABSTRACT

In this chapter the class of mimetically twinned crystals, which encompasses all ferroic materials, is discussed and primary and secondary ferroics are defined. The focus will be upon ferroelectric materials, but because of the very strong elasto-dielectric coupling in most ferroelectrics it will be necessary also to consider the ferroelastic character. In discussing past achievements in ferroelectricity the chapter draws heavily upon the earlier study by Cross and Newnham on the 'History of Ferroelectricity' [1]. Considering the build-up towards current activities, it is suggested that by the mid 1970s, a divergence began to occur between the emphasis of basic theoretical studies and the needs of the burgeoning industries using ferroelectric ceramics. This chapter considers primarily the more practical applications-driven studies which have been most active and most effective in Japan. The emphasis is upon the solid solution systems required to tune the ferroelectric phase transitions and to manipulate the intrinsic dielectric, piezoelectric, and pyroelectric properties. The newer composite technologies which we considered address the engineering of tensor properties by control of the flux paths through the solid designed by manipulation of the phase connectivity, the symmetry and the scale of the composite. Device needs driving towards the integration of ferroelectrics into more complex lamellar structures will be discussed briefly.

General trends for the future suggest: a movement towards the growth

71

of thin films of ferroelectrics, polycrystal films on silicon both for memory and capacitor applications, and single-crystal films of complex chemistry, tuned for electro-optic application and fabricated by epitaxy onto a simpler matching host. Problems of the nature and control of polar surfaces must become more important and in general the ferroic behavior of very small volumes needs to be considered. The recent demonstration of superparaelectricity and the possibility of super-paraelasticity are intriguing questions. For composites one looks to-wards the beginnings of systems in which both sensors and responders may be incorporated into the same structure so that an electronic interconnect may be used to generate a 'smart' solid—smart in the sense that it can adjust its response to the impinging field variables under command. Finally, it may be noted that in the new high-T_c superconductors the oxygen octahedron arrangement is conducive to ferroicity. The yttrium barium cuprate is certainly ferroelastic and there is mounting evidence of ferroelectric or near ferroelectric properties in several of these oxide superconductor compounds.

1. INTRODUCTION

Since the term 'ferroic' has been 'coined' from a loosely perceived analogy between the domain-related hysteretic response of the magnetization to magnetic field in iron, and somewhat similar hysteretic responses in dielectric and elastic crystals, it is perhaps useful to give a rigorous definition of ferroelectricity, then to extend this to ferroelasticity and to other mimetically twinned crystals in which the twin (domain) structures may be driven between equivalent equilibrium orientation states by other combinations of fields, the so-called secondary ferroic crystals [2, 3].

The bulk of this chapter will deal with ferroelectric crystals, but owing to the large magnitudes of piezoelectric and electrostrictive coupling, many ferroelectrics are also ferroelastic and there is need to consider the elastic interactions which play an important role in all ferroelectric phenomena. To consider the early evolution of fer-roelectricity the paper makes extensive use of data provided in an earlier study of the 'History of Ferroelectricity' by L. E. Cross and R. E. Newnham [1]. In that paper, seven decades in the development were named, covering the whole lifetime of ferroelectricity from 1920 through to 1990. Here we particularly highlight the decade of high science over the period from 1960 to 1970, which laid the foundation for our present understanding of polarization, phase transitions, domain and bulk properties of the practically useful oxide ferroelectrics. In many ways, this era was so successful that most theoretical physicists, believing the subject 'mined out', moved on to more exotic but much less useful ferroics, to problems like the incommensurate structures, the lock-in transitions, phasons and amplitudons, the elementary excitations of the incommensurate form, and so on. In the meantime, industry, particularly in Japan, was building needs for ferroelectrics with better-controlled properties for capacitor dielectrics, piezoelectrics, pyroelectrics, and electro-optic applications. In the age of diversification which followed we will follow the work upon the more practical materials.

For many purposes the essential requirement was the development of solid-solution systems based upon the single perovskite ferroelectric forms in which major solid substitution is used to bring the extrema in properties which occur near the ferroelectric transitions into usable temperature ranges. More recently this has been followed by phenomenological thermodynamic studies which seek to document the intrinsic single-domain properties of the crystal and separate domain-wall, phase-boundary and defect contributions to the measured bulk properties of these complex ceramics. Alternative composite approaches which address property optimization by using interconnection in a two-phase system to shunt fluxes and fields in the desired manner so as to optimize the tensor components in the figure of merit for piezoelectric, pyroelectric systems will be reviewed briefly.

In ceramic packaging for semiconductor chips, new miniaturized architectures lead to both higher circuit densities and higher speeds and thus the need to bring passive components closer to the chip. Integrated multicomponent multilayer technology will be discussed briefly to bring out essential features of current and potential future material needs.

For the future, several general trends are evident. Firstly, it appears that we are beginning to be able to produce proper crystalline ferroelectric oxides in thin-film form. Two different areas appear to show special promise: firstly, the application of thin ($\sim$5000 Å) polycrystal films to silicon, both for switching and memory, and for passive capacitance applications; secondly, the epitaxial growth of single crystals of complex composition upon matching but chemically simpler hosts for optical electro-optic and optical switching applications. As these thin-film approaches mature there will clearly be major needs for more detailed study of polar surfaces, and upon the effects of scale upon ferroelectric properties of one-, two-, and three-dimensional comminution. Already in ferroelectrics of complex composition there is evidence of heterophase fluctuations giving rise to superparaelectric behavior, and the possible effects of elastic fields upon the dielectric lead automatically to the intriguing possibility of superparaelastic behavior.

For composites, the chapter will explore the beginning of attempts to implant sensors and responders in the same composite structure, to interconnect them through appropriate electronics so as to architect a 'smart' material which, by changing the gain characteristic of the coupling, could adjust its response in a programmed manner to impinging dynamical fields.

Finally, the most recent additions to the cadre of ferroic solids are the $YBa_2Cu_3O_{7-\delta}$ superconductors. The supergroup:subgroup relations at the tetragonal:orthorhombic structure change (4/mmm:mmm) and the evident twin structures are clear indications of potential ferroelasticity. In this paper we will briefly review additional evidence which also suggests ferroelectric transitions in these oxides.

2. DEFINITIONS OF FERROIC MATERIALS

It has been generally accepted for a long time that the term 'ferromagnetic' be used to describe the magnetic properties of iron. Basic features of ferromagnetism are the appearance of a spontaneous magnetic moment at temperatures below a Curie point, the occurrence of domains in the material over which the magnetization is essentially uniform and the ability under the influence of an external magnetic field to modulate these domains orientations so as to leave states of high remanent magnetization (permanent magnets) at zero field. The domain reorientation in cyclic fields leads to the well-known hysteresis figure for a ferromagnet (Fig. 1).

In Rochelle salt, the appearance of a similar panoply of properties in the dielectric response led Valasek [4] to propose the term 'ferroelectric'. Looking now for a rigorous definition, it has been proposed that a ferroelectric is a material which exhibits one or more ferroelectric phases in a realizable range of pressure and temperature.

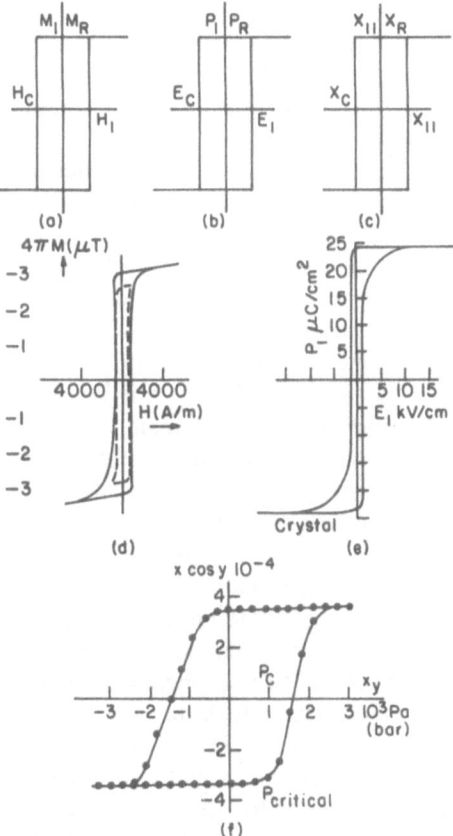

FIG. 1. Primary ferroics: (a) ferromagnetic; (b) ferroelectric; (c) ferroelastic, (d) ceramic $Co_{0.1}Fe_{0.9}Fe_2O_4$, 25°C; (e) single-crystal $BaTiO_3$, 25°C; (f) single-crystal $Pb_3P_{1.6}V_{0.4}O_8$.

In the ferroelectric phase, the crystal is spontaneously electrically polarized and the polarization has more than one possible equilibrium orientation, i.e. a domain structure. To establish ferroelectricity it must be demonstrated that the polarization can be reoriented between the different orientation (domain) states by a realizable electric field.

In ferroelectrics there are often practical difficulties and it may be hard to demonstrate unequivocally that persistent electric polarization is an equilibrium property and not just a persistent electret state. The requirement to demonstrate reorientability leads also to a conditional definition which hinges upon an often ill-defined limit, i.e. what electric field should be 'realizable' in a given insulator before breakdown. The situation becomes even more messy if the material is highly conducting, so that compensation charge completely screens the electric polarization.

It was perhaps natural that as the study of martensitic transitions progressed the elastic hysteresis observed in copper:gold, indium:thallium, and a number of oxides (Fig. 1c) should raise the question of ferroelasticity. Here the concept of spontaneous strain is a little more difficult, even when a paraelastic prototype from which the ferroelastic evolves is known; since thermal expansion (contraction) will obtrude on the dimension changes, the nature of spontaneous strain is unclear. Thus, for ferroelastics it is necessary to define the 'unstrained' state by applying to the domain all symmetry operations required to get to all other twin (domain) states and subtracting out the differential deformations. In this way an 'undeformed' state (prototype) can be unequivocally defined and spontaneous strains can be delineated. Again, to demonstrate ferroelasticity the spontaneous strain must be reorientable by a realizable elastic stress.

The definitions given above define the three simple primary ferroics: ferromagnetic, ferroelectric, and ferroelastic (Fig. 1). It is evident however, that the nub of the ferroic phenomenon is the existence of distinct twin (domain) states, which in the above cases differ in magnetization, in polarization, or in elastic strain and thus can be driven between states by magnetic, electric, or elastic stress. Carrying this thinking further [5, 6] leads to definition of six secondary ferroics (Table 1), where domains differ now in dielectric susceptibility (ferrobielectric), magnetic susceptibility (ferrobimagnetic), elastic compliance tensor (ferrobielastic), piezoelectric tensor (ferroelastoelectric), piezomagnetic tensor (ferromagnetoelastic), and magnetoelectric tensor (ferromagnetoelectric). In these cases, again, to demonstrate ferroicity the twins must be reorientable under the appropriate combination of driving fields.

After a flurry of interest in the 1970s, activity upon secondary ferroics declined, owing to the very high drive levels often needed to effect secondary domain switching. With new interest and activity in very thin-film ferroelectrics, where breakdown fields are improved by orders of magnitude, one may perhaps expect to see new interest in these secondary ferroics.

TABLE 1
Primary and secondary ferroics

Ferroic class	Orientation states differ in	Switching force	Example
Primary			
Ferroelectric	Spontaneous polarization	Electric field	$BaTiO_3$
Ferroelastic	Spontaneous strain	Mechanical stress	β-$Au_xCu_{1-x}Zn$
Ferromagnetic	Spontaneous magnetization	Magnetic field	Fe_3O_4
Secondary			
Ferrobielectric	Dielectric susceptibility	Electric field	$SrTiO_3(?)$
Ferrobimagnetic	Magnetic susceptibility	Magnetic field	NiO
Ferrobielastic	Elastic compliance	Mechanical stress	SiO_2
Ferroelastoelectric	Piezoelectric coefficients	Electric field and mechanical stress	$NH_4Cl(?)$
Ferromagnetoelastic	Piezomagnetic coefficients	Magnetic field and mechanical stress	CoF_2
Ferromagnetoelectric	Magnetoelectric coefficients	Magnetic field and electric field	Cr_2O_3

3. FERROELECTRIC PAST

In a recent study, Cross and Newnham [1] have traced the evolution of ferroelectricity through seven decades (Table 2). From that study, we pick out the fiducial dates which are of most interest for this paper.

In the early Rochelle salt period (Table 3) the first analogy to ferromagnetism by J. Valasek [4], the first practical application of a ferroelectric to bimorph 'benders' and 'twisters' and their early use in tube-type phonographs by C. B. Sawyer [7], and the elegant phenomenological description by H. Mueller [8] later applied by A. F.

TABLE 2
Important events in ferroelectricity

1920–1930	Rochelle salt period: discovery of ferroelectricity
1930–1940	KDP age: thermodynamic and atomistic models of ferroelectricity
1940–1950	Early barium titanate era: high-K capacitors developed
1950–1960	Period of proliferation: many new ferroelectrics discovered
1960–1970	Age of high science: soft modes and order parameters
1970–1980	Age of diversification: ferroics, electro-optics, thermistors
1980–1990	Age of integration: packages, composites, and integrated optics
1990–2000	Age of miniaturization: size effects, manipulated modes and dipoles

TABLE 3
Early Rochelle salt period

1655	Seignette	La Rochelle
		First recorded synthesis of
		Rochelle Salt
1880	Curie,	Piezoelectric properties
	Curie	
1894	Pockels	Dielectric anomalies, Kerr electro-optical
		effect
1917	Anderson,	Piezoelectric effects practical
	Nicolson,	applications
	Cady	
1921	Valasek	Analogy to magnetism
to		
1924	Swann	Origin of ferroelectricity
1930	Sawyer	Rochelle salt bimorphs,
		'benders' and 'twisters'
1937	Jaffe	Symmetry change at T_c
1940	Mueller	First complete phenomenological
		theory

Devonshire [9] to $BaTiO_3$ are of most importance. It is interesting to note that Mueller appreciated quite clearly that Rochelle salt could be ferroelastic with secondary ferroelectricity, or that the instability could be just piezoelectric, and devised experimental tests to distinguish these cases.

The KDP age (Table 4) yielded a second much-needed ferroelectric of much simpler structure, the first quantitative atomistic models and the first crystal with large longitudinal electro-optic coefficients.

In the 1940s (Table 5) the discovery of ferroelectricity in the simple perovskite $BaTiO_3$ gave a 'shot in the arm' to the whole subject—a

TABLE 4
Potassium dihydrogen phosphate group

1930	West	Crystal structure
1935	Busch,	First report of ferroelectricity
	Scherrer	
1941	Slater	Proton ordering theory
1953	Bacon,	Neutron diffraction locating
	Pease	hydrogen
1944	Zwicker,	Large longitudinal electro-optical
	Scherrer	effect

TABLE 5
Perovskite era (early $BaTiO_3$ period)

1940–	Wainer and Salomon	Discovery of $BaTiO_3$
1943	Ogawa	
	Wul and Goldman	
1945	Gray	First operating poled $BaTiO_3$ transducer
1945	von Hippel	Ferroelectricity in $BaTiO_3$
1946	Ginsburg	
1946	Megaw	Crystal structure
1947	Matthias and Merz	Single crystals
1948	Matthias, von Hippel, Blattner, Kanzig, Merz, Sutter, Cross, Dennison, and Nicholson	Optical domain structure, temperature and field effects
1949	Devonshire	Phenomenology
1949	Kay, Vousden	Structure changes
1952	Merz	Single crystal properties
1954	Merz	Switching studies
1953	Fraser, Danner, Pepinsky	Neutron diffraction study of $BaTiO_3$

material which was structurally simple, cubic above T_c, optically transparent with easily distinguished domains but encompassing three ferroelectric phases between 408 K and 4 K.

Over the next 10 years which were perhaps the most exciting in the history of the subject (Table 6) some 24 new ferroelectrics were discovered. No doubt, the friendly competition between Ray Pepinsky and his group at Penn State and Berndt Matthias and co-workers at Bell Labs stimulated much of this activity, but it was during this time that most of today's practical materials, $LiNbO_3$, $KNbO_3$, $PbNb_2O_6$, SbSI, and many others were discovered.

Probably, however, the 10 years from 1960 to 1970, which we termed the age of high science really laid the foundation for the present understanding of ferroelectrics. The formulation of the soft-mode theory by Cochran and Anderson [10, 11] came just at the time when inelastic neutron scattering and Raman spectroscopy were being widely applied, tools able to confirm the dynamical phenomenology for $SrTiO_3$, $KTaO_3$, and other perovskites. Over the period, the complete symmetry classification for all ferroics was effected by Aizu and Shuvalov [12–15]. The first improper ferroelectric ($Gd_2(MoO_4)_3$) was discovered and the basis of ferroelectric behavior in $PbTiO_3$ and in

TABLE 6
Period of proliferation

1949– 1960	20 Perovskite compounds.	Matthias Smolenskii	WO_3 to $Pb_3MgNb_2O_9$
1949	$LiNbO_3$ family	Matthias	
1951	$LiTlC_4H_4O_6 \cdot H_2O$	Matthias	
1952	$Cd_2Nb_2O_7$ pyrochlore family	Cook, Jaffe	
1953	$PbNb_2O_6$ tungsten bronze	Goodman	
1955	GASH family	Holden	
1956	$Sn(NH_2)_2$ thiourea	Soloman	
1956	TGS family	Matthias	
1956	$(NH_4)_2SO_4$ family	Matthias	
1956	Colemanite	Goldsmith	
1956	$(NH_4)_2Cd_2(SO_4)_3$ family	Jona, Pepinsky	
1957	Alums	Jona	
1957	Dicalcium strontium propionate	Matthias	
1957	Boracite family	Le Corre	
1958	$(NH_4)HSO_4$	Pepinsky	
1958	$NaNO_2$ family	Sawada	
1958	KNO_3	Sawada	
1959	$LiH_3(SeO_3)_2$ family	Pepinsky	
1959	$(NH_4)NaSO_4$	Pepinsky	
1960	$N(CH_3)_4 \cdot HgCl_3$ family	Fatuzzo, Merz	
1960	$K_4Fe(CN)_6 \cdot 3H_2O$ family	Waku	
1962	SbSI family	Fatuzzo	
1963	$YMnO_3$ family	Bertaut	

the technologically important $PbTiO_3 : PbZrO_3$ solid solutions was established.

About the mid 1970s it became rather clear that the basic physical interest in ferroelectricity was diminishing but that the technological needs of major industries based upon ferroelectrics in oxide ceramic form were increasing rapidly. Thus, while theoreticians rode off to tilt at the windmills of more esoteric incommensurate transitions, lock-in behavior, phasons and amplitudons, and other good things, electroceramists were forced to address the more practical problems of making useful dielectric, piezoelectric, pyroelectric, and electro-optic materials from these ferroelectric families. It is this more practical evolution which we wish to trace to the present and to project into the future.

4. FERROELECTRICS PRESENT

In a simple cubic perovskite such as $BaTiO_3$ (Fig. 2), it is evident that the dielectric response goes through sharp maxima near the

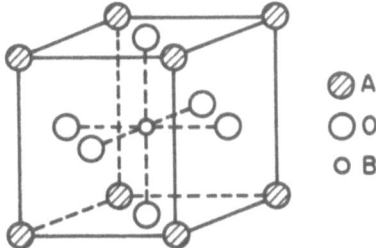

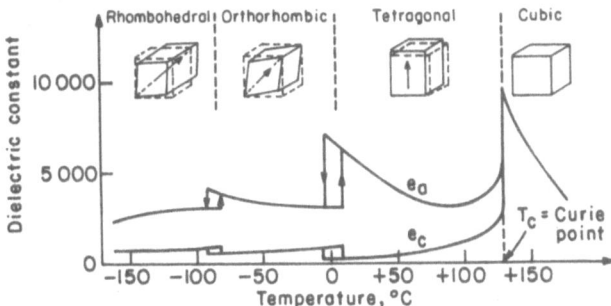

⊘ A
○ O
○ B

Fig. 2. Cubic m3m prototype structure of perovskite type ABO_3 compound. For $BaTiO_3$, $A = Ba^{2+}$, $B = Ti^{4+}$, $O = O^{2-}$.

paraelectric–ferroelectric and the ferroelectric–ferroelectric phase transitions (Fig. 3). It was appreciated quite early that these phase transitions could be moved along the temperature axis by solid solution. The types of change produced are evidenced in Fig. 4, which delineates the effects of simple A-site and B-site substitutions on the ABO_3 structure. It may be noted that the process is not one of 'doping' but of major substitution in the range 10–30 mole% to effect these lattice-controlled properties. As might be expected, the influence on the lower-temperature ferroelectric–ferroelectric transitions is more complex and the influence on the total dielectric response, which includes domain and defect contributions, is even more complex. In $BaTiO_3$ the developments of modified dielectric for MLC applications has been largely empirical, and even the grain size effect in high-purity single-phase material is still in dispute. For the newer relaxor ferroelectric formulations based on lead magnesium niobate and other complex perovskites, the mechanisms are perhaps a little better understood, as will be discussed later.

Fig. 3. Dielectric permittivity of single-crystal $BaTiO_3$ as function of temperature for weak fields.

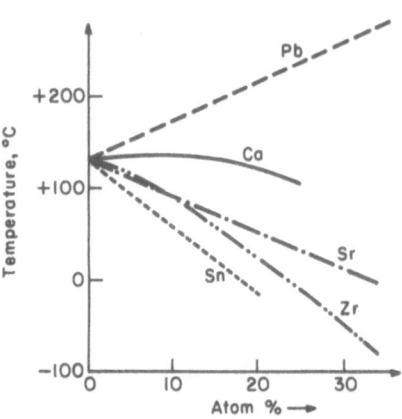

Fɪɢ. 4. Shift of the Curie temperature in $BaTiO_3$ by substitution of Ba^{2+} ions with Pb^{2+}, Ca^{2+}, or Sr^{2+} ions or of Ti^{4+} ions by Zr^{4+} or Sn^{4+} ions.

For the $PbZrO_3 : PbTiO_3$ solid solution system almost universally used in piezoelectric applications, again the development leading to the morphotropic phase boundary compositions was largely empirical. However, now the system is being more effectively delineated. Haun and co-workers [16, 17] at Penn State have developed a full elastic Gibbs function for the whole PZT family and have painstakingly built up the full family of constants and their composition and temperature dependence needed to describe all phases throughout the whole phase space. The equation for ΔG is of course complicated (Fig. 5) but does permit a complete calculation of the boundaries of all phases, in excellent agreement with the measured phase stability (Fig. 6), and of the temperature dependence of the averaged intrinsic single-domain properties for any composition. As an example, the dielectric permittivity of the 52:48 PZT is calculated as a function of temperature and compared with experimental data for Navy type I to V compositions (Fig. 7). The excellent agreement for these and for other Zr : Ti ratio ceramics at 4 K where all extrinsic effects are frozen out gives confidence in the predictive capability of the phenomenology and underscores the major importance of domain- and defect-controlled contributions to the total response.

An alternative approach to tuning the properties of materials for specific applications is exemplified by the polymer–ceramic composite studies of Newnham and co-workers [18]. Examination of the different

$$\Delta G = \alpha_1[P_1^2 + P_2^2 + P_3^2] + \alpha_{11}[P_1^4 + P_2^4 + P_3^4]$$
$$+ \alpha_{12}[P_1^2P_2^2 + P_2^2P_3^2 + P_3^2P_1^2] + \alpha_{111}[P_1^6 + P_2^6 + P_3^6]$$
$$+ \alpha_{112}[P_1^4(P_2^2 + P_3^2) + P_2^4(P_1^2 + P_3^2) + P_3^4(P_1^2 + P_2^2)]$$
$$+ \alpha_{123}P_1^2P_2^2P_3^2 + \sigma_1[p_1^2 + p_2^2 + p_3^2] + \sigma_{11}[p_1^4 + p_2^4 + p_3^4]$$
$$+ \sigma_{12}[p_1^2p_2^2 + p_2^2p_3^2 + p_3^2p_1^2] + \sigma_{111}[p_1^6 + p_2^6 + p_3^6]$$
$$+ \sigma_{112}[p_1^4(p_2^2 + p_3^2) + p_2^4(p_1^2 + p_3^2) + p_3^4(p_1^2 + p_2^2)]$$
$$+ \sigma_{123}p_1^2p_2^2p_3^2 + \mu_{11}[P_1^2p_1^2 + P_2^2p_2^2 + P_3^2p_3^2]$$
$$+ \mu_{12}[P_1^2(p_2^2 + p_3^2) + P_2^2(p_1^2 + p_3^2) + P_3^2(p_1^2 + p_2^2)]$$
$$+ \mu_{44}[P_1P_2p_1p_2 + P_2P_3p_2p_3 + P_3P_1p_3p_1] + \beta_1[\theta_1^2 + \theta_2^2 + \theta_3^2]$$
$$+ \beta_{11}[\theta_1^4 + \theta_2^4 + \theta_3^4] + \gamma_{11}[P_1^2\theta_1^2 + P_2^2\theta_2^2 + P_3^2\theta_3^2]$$
$$+ \gamma_{12}[P_1^2(\theta_1^2 + \theta_2^2) + P_2^2(\theta_1^2 + \theta_3^2) + P_3^2(\theta_1^2 + \theta_2^2)]$$
$$+ \gamma_{44}[P_1P_2\theta_1\theta_2 + P_2P_3\theta_2\theta_3 + P_3P_1\theta_3\theta_1]$$
$$- \tfrac{1}{2}S_{11}[X_1^2 + X_2^2 + X_3^2] - S_{12}[X_1X_2 + X_2X_3 + X_3X_1]$$
$$- \tfrac{1}{2}S_{44}[X_4^2 + X_6^2 + X_6^2] - Q_{11}[X_1P_1^2 + X_2P_2^2 + X_3P_3^2]$$
$$- Q_{12}[X_1(P_2^2 + P_3^2) + X_2(P_1^2 + P_3^2) + X_3(P_1^2 + P_2^2)]$$
$$- Q_{44}[X_4P_2P_3 + X_5P_1P_3 + X_6P_1P_2] - Z_{11}[X_1p_1^2 + X_2p_2^2 + X_3p_3^2]$$
$$- Z_{12}[X_1(p_2^2 + p_3^2) + X_2(p_1^2 + p_3^2) + X_3(p_1^2 + p_2^2)]$$
$$- Z_{44}[X_4p_2p_3 + X_5p_1p_3 + X_6p_1p_2] - R_{11}[X_1\theta_1^2 + X_2\theta_2^2 + X_3\theta_3^2]$$
$$- R_{12}[X_1(\theta_2^2 + \theta_3^2) + X_2(\theta_1^2 + \theta_3^2) + X_3(\theta_1^2 + \theta_2^2)]$$
$$- R_{44}[X_4\theta_2\theta_3 + X_5\theta_1\theta_3 + X_6\theta_1\theta_2]$$

FIG. 5. Expansion of the elastic Gibbs function for a crystal with m3m prototype symmetry including possibility for both ferroelectric and antiferroelectric phases and of tilts of the oxygen octahedral network.

materials figures of merit for several piezoelectric, pyroelectric, and electro-optic devices indicates clearly the need to be able to control the influence of the different tensor components. For composites, Newnham has developed a set of simple connectivity models based on stacking cubes (Fig. 8) which show how flux paths can be limited by the choice of geometry. Making use of these concepts and the resulting connectivity notation, a whole range of macro composites (Fig. 9) has been tested out for large-area hydrophone application. Particularly, the type with (1113) connectivity using a transversely reinforced rod structure turns out to be exceedingly efficient (Fig. 10), with a figure of merit $d_h g_h$ more than 1000 times that of monolithic PZT.

The advantage of the polymer–ceramic composite approach is that

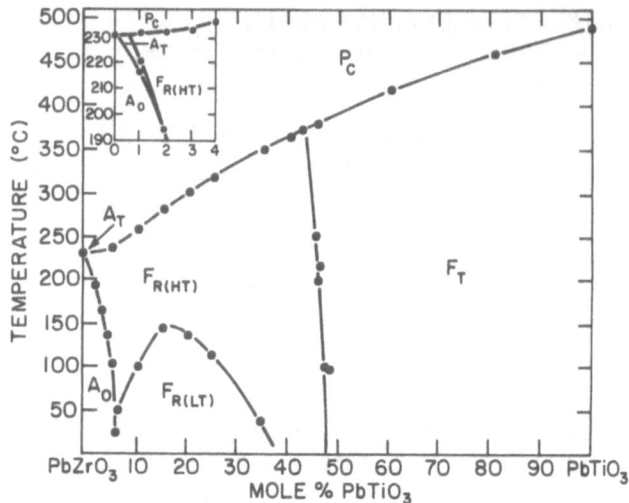

FIG. 6. Fitting between experimental and theoretically derived phase diagram for $PbZrO_3$: $PbTiO_3$ solid solutions.

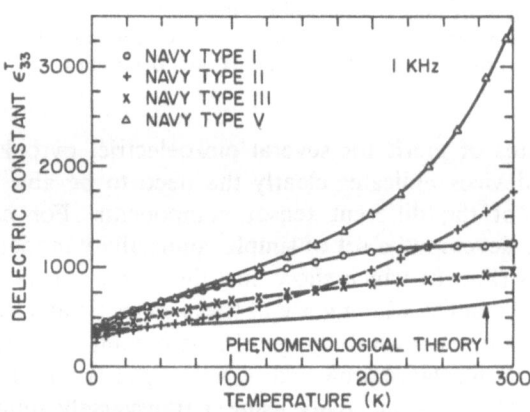

FIG. 7. Comparison of calculated and measured permittivity for PZT of the 52:48 Zr:Ti composition. Experimental samples are 'doped' to Navy Specification for type I, II, III, and IV.

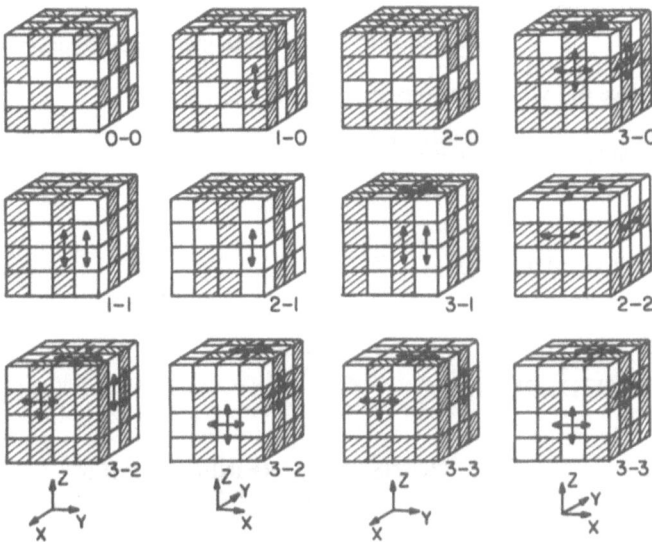

FIG. 8. Simple cubes model to illustrate the possible phase interconnection in a simple two-phase system.

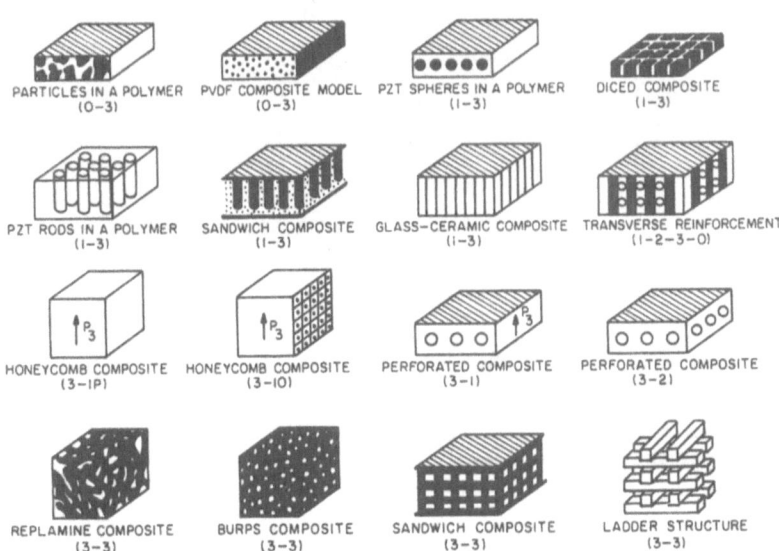

FIG. 9. Examples of PZT:polymer composites fabricated to test connectivity models.

NAVY HYDROPHONE

Up to 1975   Material lead zirconate titanate piezoelectric ceramic PZT
Power figure of merit $d_h g_h$
  Produce of hydrostatic voltage × hydrostatic charge

$d_h g_h$ in tensor form $\rightarrow \dfrac{(d_{333} + 2d_{311})^2}{\varepsilon_{33}}$

$$\boxed{d_h g_h \sim 100 \cdot 10^{-15} \ \text{m}^2/\text{Newton}}$$

PROBLEM $d_{333} \cong -2d_{311}$ ε_{33} very large

COMPOSITE SOLUTION

TRANSVERSE REINFORCEMENT
(1–2–3–0)

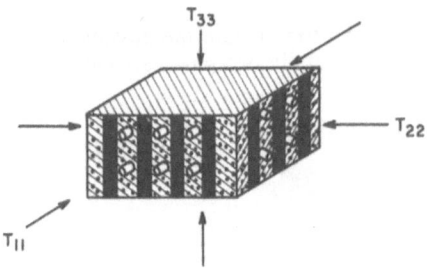

$$p = \begin{bmatrix} -T_{11} & 0 & 0 \\ 0 & -T_{22} & 0 \\ 0 & 0 & -T_{33} \end{bmatrix}$$

$\left.\begin{matrix} \\ \\ \end{matrix}\right\}$ Taken up on transverse reinforcement

Enhanced on PZT
Polymer acts like a tent

ε_{33} much reduced by the polymer
Composite 10 v% PZT 90% rubber

$$\boxed{d_h g_h \cong 150\,000 \cdot 10^{-15} \ \text{m}^2/\text{Newton}}$$

FIG. 10. A (1113) composite using PZT and glass fibers in a polymer matrix, which enhances the hydrophone figure of merit $d_h g_h$ by a factor of more than 1000.

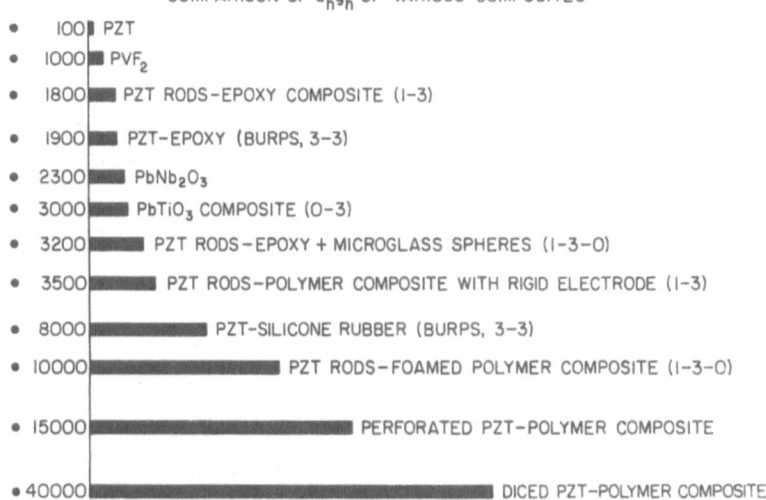

FIG. 11. Figures of merit for large-area hydrophones which can be developed using various composite geometries.

by choosing volume fraction and appropriate connectivity, a wide range of properties (Fig. 11) can be developed from the same two components.

Key elements for the composites are listed in Table 7. Most interest, so far, has been focused upon connectivity and symmetry, but recently Smith [19] has shown that by appropriate choice of scale, the composites can be used to significant advantage in high-frequency electromedical acoustic tomography (Fig. 12).

Another area of high current interest for the practical application of ferroelectrics is in the packaging of semiconductor chips. A major fraction of the market for MLC capacitors has been in association with semiconductor chips in digital logic and memory applications. With the rapid evolution of the semiconductor industry towards smaller feature sizes, higher densities, and faster clock rates, the need to bring the capacitor even closer to minimize connection inductance becomes paramount. One possible solution is to put the capacitor, along with other passive components into the package itself and the Multilayer Monolithic Multicomponent ceramic chip carrier from NEC is an interesting contender. Advantages of the system are listed in Table 8.

TABLE 7
The age of 'high' science

1959	Cochran	The soft mode
1960	Anderson	Description of $BaTiO_3$
1962	Cowley	Confirmation in $SrTiO_3$ by inelastic neutron scattering
1962	Barker, Tinkham, Spitzer, Miller, Kleinman, Howarth	Confirmation in $BaTiO_3$, $SrTiO_3$ by IR reflectance
1963	Miller Miller, Kleinman, Savage	Optical SHG KDP, $BaTiO_3$
1967	Kaminow, Damon	Raman spectra in KDP
1967	Fleury, Worlock	Raman spectra in $SrTiO_3$, $KTaO_3$
1967	Johnston, Kaminow	Raman spectra in $LiNbO_3$, $LiTaO_3$
1968	Fleury	Soft modes and 106 K transition in $SrTiO_3$
1968	Cross, Fouskova, Cummins	'Peculiar' ferroelectricity in $Gd_2(MoO_4)_3$
1967–1969	Aizu Shuvalov	Complete symmetry classification of all ferroelectrics
1970	Pytte Sanikov Aizu	Improper ferroelectricity in $Gd_2(MoO_4)_3$
1970	Shirane	Inelastic neutron $PbTiO_3$
1971	Nunes	Studies of soft $KNbO_3$
1971	Harada	Modes in $BaTiO_3$

The make-up is delineated schematically in Fig. 13, and the physical properties of the lead borosilicate glass-bonded alumina of the support structure are given in Table 9.

To permit co-firing of the chemically different species, temperature and shrinkage have been brought together and in Table 10 this is contrasted with the widely disparate processing conditions of the components used in non-monolithic approaches. The characteristics of the low-firing high-K lead-based relaxor dielectric used in the capacitor section is given in Fig. 14.

5. FERROELECTRICS: FUTURE PROSPECTS

Predicting the future in any area of research which is driven by technology needs is always difficult and subject to wide margins of

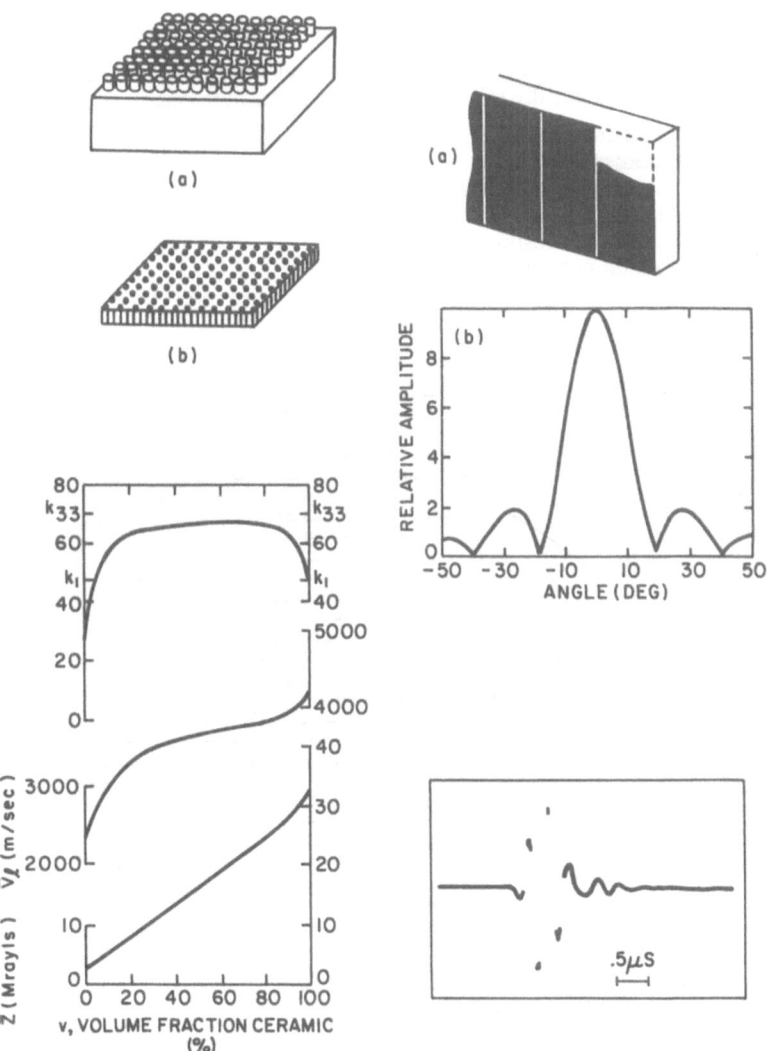

FIG. 12. Piezoelectric composites for medical ultrasonic imaging: (a) diced ceramic PZT; (b) back filled with polymer and separated from the base [19].

TABLE 8
Monolithic multi-components ceramic technology after NEC

Lead borosilicate glass bonded alumina + relaxor ferroelectric substrate

The new MMC substrate has three-dimensional capacitor and resistor elements inside it and has the following excellent characteristics:

1. Silver–palladium, silver, and gold can be used in fabricating wiring or circuit patterns.
2. A large capacitance capacitor can be formed in the substrate (10 pF–3 μF).
3. A plurality of resistor and capacitor elements can be three-dimensionally constructed in the substrate.
4. Miniaturization and low cost can be achieved.

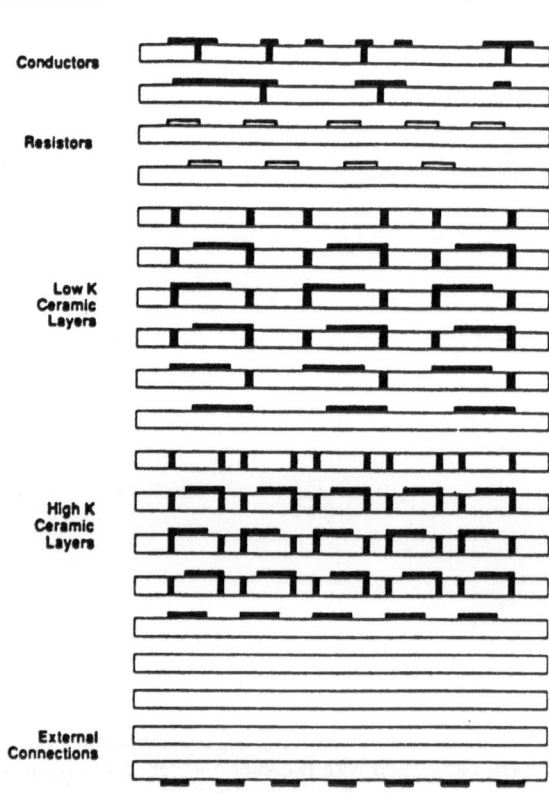

FIG. 13. Monolithic multicomponent ceramic in schematic form, illustrating the make-up of the different co-fired layers.

TABLE 9
Physical characteristics of the NEC insulator

Flexural strength	$3\,000\,kg/cm^3$
Thermal expansion coefficient	$42 \times 10^{-7}\,deg^{-1}$
Dielectric constant (1 kHz)	7·5
Dissipation factor (1 kHz)	0·3%
Insulation resistance (50V d.c.)	$2·9 \times 10^{14}\,\Omega$. cm
Leak current (20V d.c.)	$<1\,\mu A$
Thermal conductivity	0·01 cal/deg . cm . s
Undulation	$30\,\mu m/100\,mm$
Roughness	$0·3\,\mu m$ Ra
Shrinkage	12·5%

error. However, with that caveat in mind it would appear that:

(1) There are several technology areas driving a need for thin ferroelectric crystalline films.
(2) Thin film and composite technologies will require a better understanding of polar surfaces, and of the changes which occur in the surface and near the surface during ferroelectric switching.
(3) Both the above areas will require study of the effects of scale on the ferroelectric phenomena, where the scale changes may be one-, two-, or three-dimensional.
(4) There is already considerable evidence that the compositionally heterogeneous relaxor ferroelectrics are in mixed-phase structures and that the scale of the heterophase fluctuations is such

TABLE 10
Firing temperatures for cofire and for 'normal' components

	New system		Usual system	
	Material	*Sintering Temperature (°C)*	*Material*	*Sintering Temperature (°C)*
Substrate	Glass–ceramic	900	Al_2O_3	1 600
Dielectric ceramic	Low-firing ceramic	900	$BaTiO_3$	1 400
Resistor	RuO_2	900	RuO_2	900
Conductor	Au,Ag–Pd	900	W,Mo	1 600
			Au,Ag–Pd	900

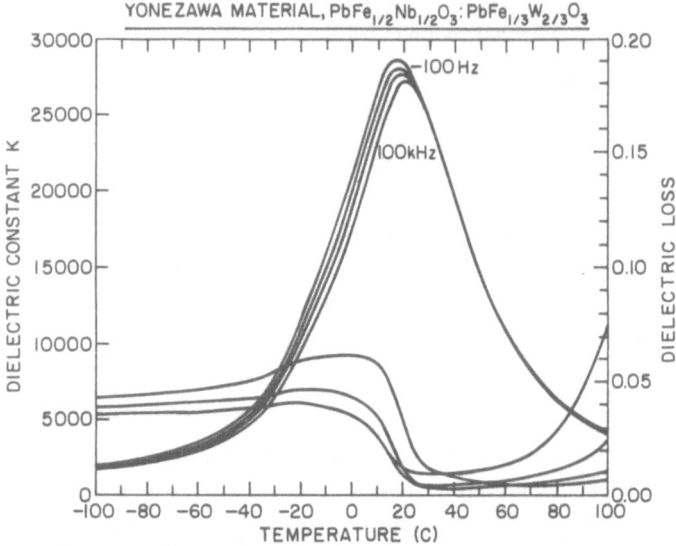

FIG. 14. Dielectric formulation similar to that used for the high-K dielectric in the monolithic multicomponent ceramic.

that dynamical disordering occurs over a wide temperature range as in the superparamagnets. Since ferroelectricity and ferroelasticity are intimately connected in perovskite ferroelectrics, one may speculate that these solids should be simultaneously superparaelastic.

(5) In composite technology, the next obvious step is to incorporate sensors and responders into the structure so as to generate 'smart' materials in which the feedback may be electronically controlled to suit the response to the ambient requirement.

(6) It is clear from the symmetry relations that the $YBa_2Cu_3O_{7-\delta}$ superconductors are probably ferroelastic in the species 4/mmmFmmm. For these oxygen octahedron structures there is now mounting evidence that the crystals are also ferroelectric.

5.1. Thin Films of Ferroelectrics

In polycrystal films, needs come from two areas. (a) In MLCs it is clear that as voltages go down in semiconductor logic and space becomes steadily more valuable, the dielectric thickness in MLCs must shrink. In present tape-cast and screen-printed capacitors, $12 \cdot 5 \, \mu m$

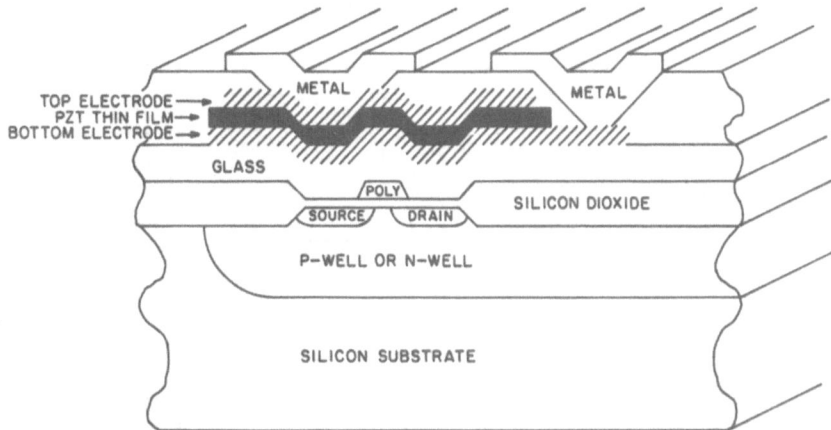

FIG. 15. PZT sputtered film on a CMOS silicon circuit, illustrative of Ramtron technology.

(1/2 mil) is about the limit. For sol–gel, spun-on or sputtered films, however, thicknesses down to 1 μm and below are achievable. At these thicknesses, however, 1 volt applied leads to fields of 10 kV/cm, which would engender high hysteretic loss in current ferroelectrics. It is probable that for these new needs a new generation of antiferroelectric formulations will be required. (b) For possible application to semiconductor logic, switchable ferroelectric films are required. Figure 15 shows a proposed PZT on CMOS configuration from Ramtron, where the ferroelectric film is 0·5 μm thick. Characteristics for such a film are shown in Fig. 16, which also lists ideal and presently available characteristics.

Elements providing archival memory on silicon of the type proposed could find wide application in credit card, keying, and security applications.

For optical applications, monocrystal films will probably be required if grain-boundary scattering is to be avoided. For many optical devices, only a single surface is required, so that epitaxy of a complex composition onto a simpler single crystal host appears attractive. R. R. Neurgaonkar at Rockwell has been exploring several promising tungsten-bronze composites (Fig. 17) which show good combinations of dielectric piezoelectric and electro-optic coefficients. Strontium barium niobates have been grown with large diameter and excellent

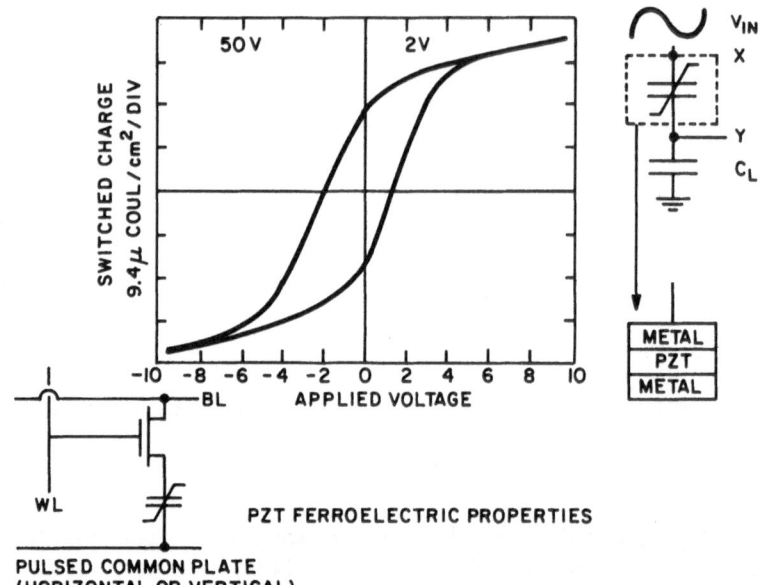

FIG. 16. Typical hysteresis loop for a PZT $0.5\,\mu m$ film: comparison of observed and 'ideal' properties for switching applications.

	Ideal	*Current*
Polarization retention	>10 years	TBD
Endurance (fatigue)	>10^{15} Cycles	>10^9 Cycles
Breakdown	>10 V	>30 V
Coercive voltage	<5 V	1 to 2 V
Net polarization charge	1 to 20 $\mu C/cm^2$	5 to 20 $\mu C/cm^2$
Curie point	>200°C	>300°C
Switching speed	1 to 5 ns	5 to 20 ns
Process compatibility	CMOS/bipolar/GaAs	CMOS
Radiation hardness	100 m rad, 10^{12} rad/sec	>10 m rad, >10^{11} rad/sec.

optical quality (Fig. 18) and more complex bronze compositions have been sputter-deposited onto the SBN substrate to yield single-crystal epitaxial films of high quality (Fig. 19).

Applications for such films will include fast T1R switches, guiding structures and switchable interconnects. Because of the excellent dielectric strength, switching times in the 1–5 ns range can be predicted.

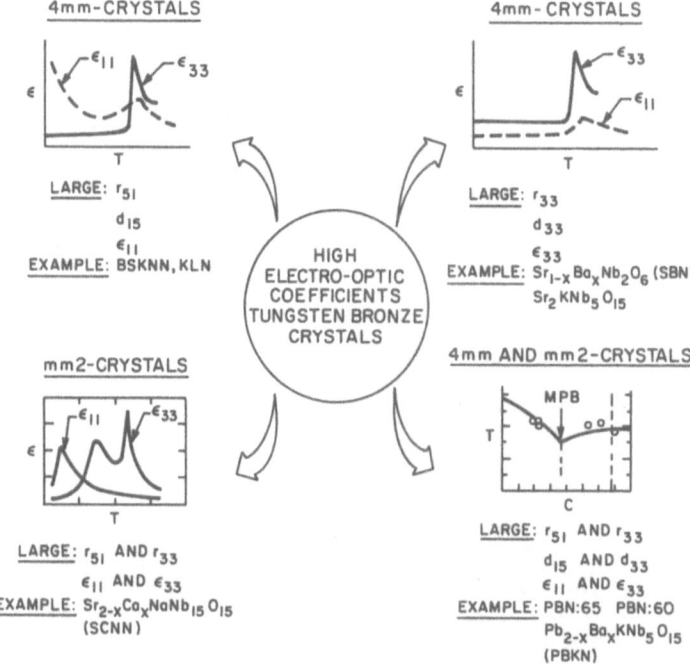

FIG. 17. Promising types of tungsten bronze structure compositions being developed by R. R. Neurgaonkar and co-workers at Rockwell.

5.2. Ferroelectric Surfaces

As the thickness of ferroelectric elements is decreased, the importance of impedance mismatch at the surface will increase. At present, the understanding of how the polar structure terminates at different electrodes is fragile and there is a real need for new work in this area. From studies on grain-grown PLZTs which are only a single crystal thick, it is evident that there are unusual surface-layer effects which are incompletely understood.

5.3. Scaling of Ferroelectric Properties

For comminuted powders and for sol–gel chemically co-precipitated and hydrothermally prepared powders, current studies upon $PbTiO_3$ suggest that neither the ferroelectric Curie temperature, nor the polarization induced c/a ratio is significantly perturbed down to sizes of ~200 Å. The scale has been checked both by BET surface analysis

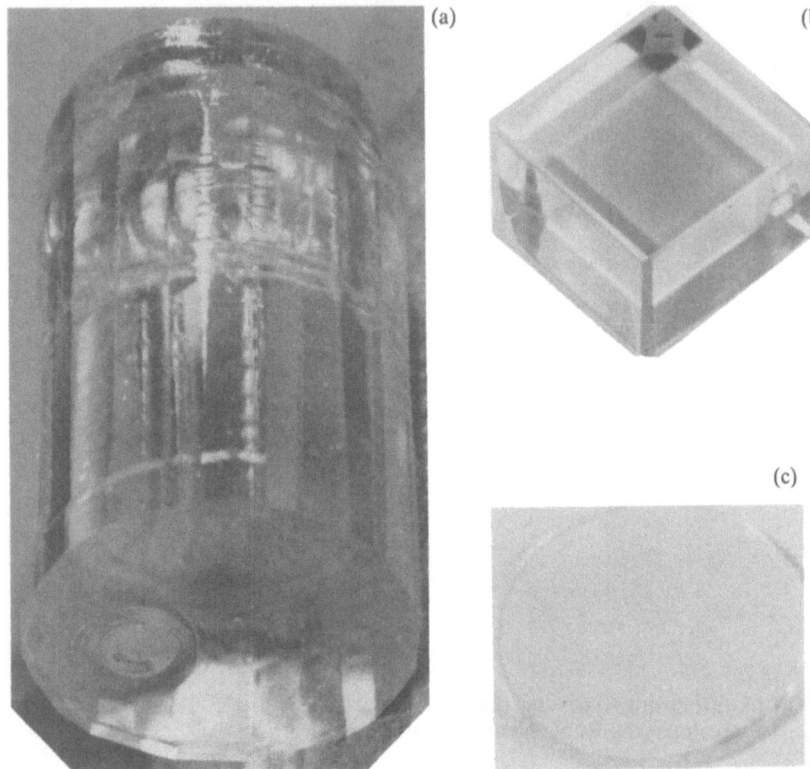

FIG. 18. Single-crystal compositions of optical quality which could be used as host surfaces for epitaxial crystal growth. (a) 3 cm diameter; (b) 2 × 2 × 2 cm cube, (c) 3 cm disk.

and by X-ray line broadening with comparable results. That the data, though highly surprising, are meaningful is confirmed by the line splitting due to tetragonality occurring on the size-broadened lines.

5.4. Superparaelectric and Superparaelastic Behavior
In PLZT on the other hand, at the 8:65:35 composition, new TEM studies confirm the micropolar character of the relaxor phase, which can be preserved down to low temperature, but then by poling with the beam current can be built up into a stable macrodomain state. Clear evidence now exists from a very wide range of studies that the relaxor ferroelectrics are true analogs of the superparamagnets.

Applications
- Guided wave optics (r_{ij}, ε)
- Spatial light modulators (r_{ij})
- 3-D storage and display ($n^3 r_{ij}/\varepsilon$)
- Pyroelectric detectors (P/ε)
- Piezoelectric sensors (d_{33}, ε)

Advantages
1. Wide range of compositions
   - Pb^{2+}-containing ferroelectrics
   - High figures-of-merit
2. Integration with substrates
   - Ferroelectrics and paraelectrics
   - Semiconductor and superconductors
3. Quality
   - Surface smoothness
   - Defect-free

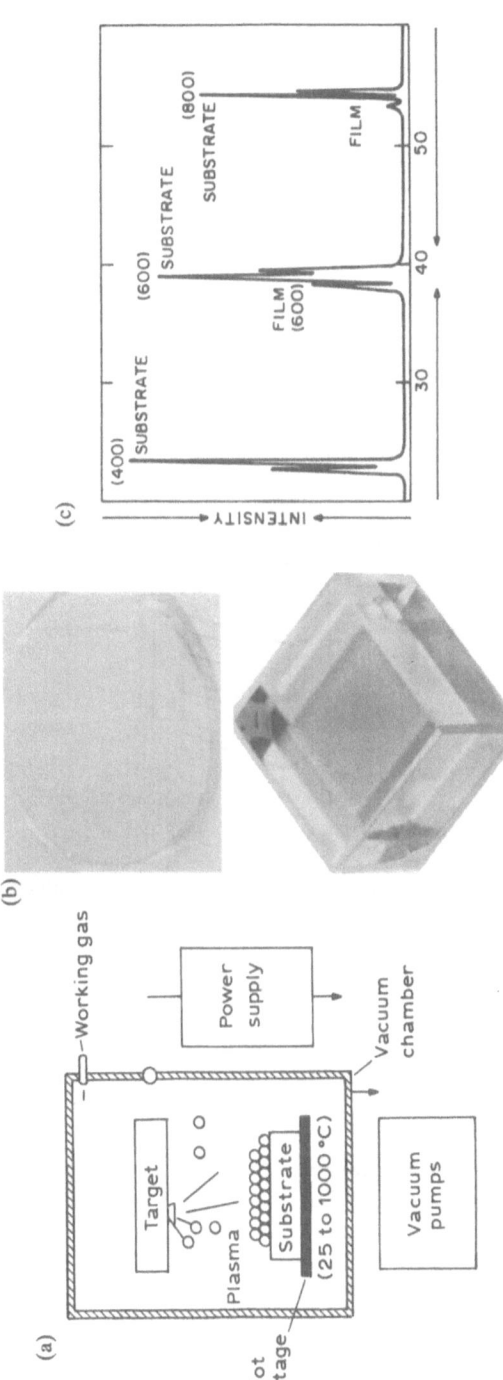

FIG. 19. Example of the growth of lead barium niobate on barium strontium niobate. The 60:40 PBN is of interest as a morphotropic phase boundary composition. (a) sputtering units; SBN: 60—cube and wafer; (c) PBN: 60 film on SBN: 60.

Current work at Penn State on the non-linear elastic behavior of these relaxors seeks to confirm superparaelasticity in these crystals.

5.5. Smart Solids

A logical extension of composite technology is the incorporation of sensors and responders with coupled electronic feedback to control properties.

A simple example for the control of the effective elastic compliance s_{11} at low frequency is shown schematically in Fig. 20. The element sketched comprises a piezoelectric stress sensor and a small feedback amplifier controlling a multilayer actuator, to give active compliance control. If, when the sensor detects increasing stress, the feedback

FEEDBACK CONTROL OF ELASTIC COMPLIANCE

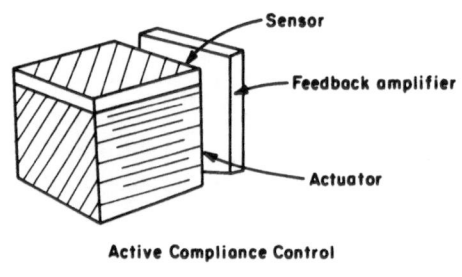

Active Compliance Control

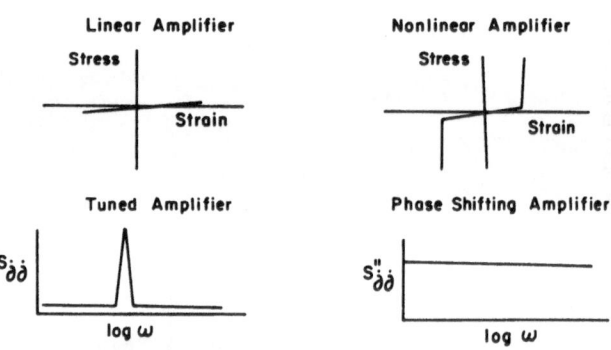

FIG. 20. Sensor–actuator combinations for the feedback control of elastic compliance. Sensor detects increasing stress. Feedback amplifier reduces length of the piezoelectric stack . . . increasing strain combination mimics an ultra soft solid.

amplifier reduces the length of the actuator, increasing the effective strain in the stack, the combination will mimic the elastic response of a soft material whose compliance can be controlled by the feedback gain.

Now since the elastic response is under electronic control it can be tuned to specific frequencies, made deliberately non-linear, or phase-shifted to mimic high-loss tangent or strong acoustic absorption.

The model proposed is a very simple example of the general principle of active composites, and the same thinking is applicable to a very wide range of possible system applications.

5.6. Ferroic Superconductors

For the $YBa_2Cu_3O_{7-\delta}$ composition (Fig. 21) it is well known that to achieve superconductivity at $90\,K$, the crystal must be annealed in oxygen to achieve an orthorhombic twinned modification (Fig. 22) of the high-temperature tetragonal form. The group-theoretical relation is exactly that expected for the ferroelastic species 4/mmmFmmm, but as yet there has been no confirmation of twin motion under stress.

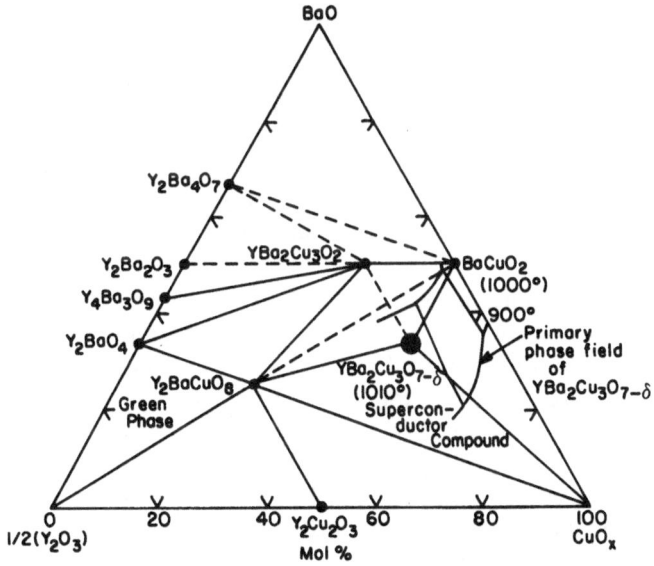

FIG. 21. Location of the $YBa_2Cu_3O_{7-\delta}$ superconductor composition in the $CuO:Y_2O_3:BaO$ phase diagram.

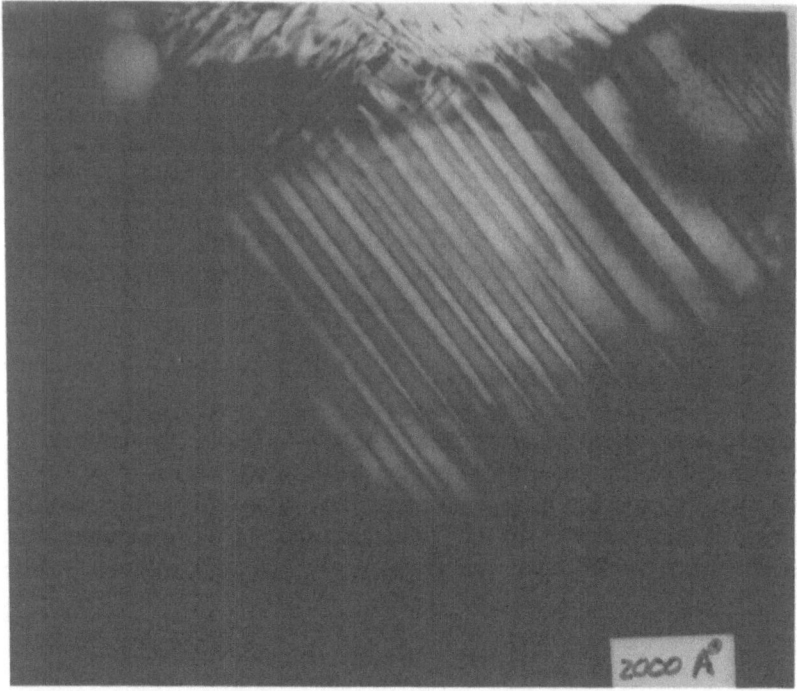

FIG. 22. Typical transmission electron microscope (TEM) picture of ferroelectric twinning in the $YBa_2Cu_3O_{7-\delta}$ superconductor.

Horowitz, Barsch, and Krumhansl [20] suggest that motion of the twin
walls   themselves   (dyadons)   may   be   important   for   the
superconductivity.

More recent studies by Kurtz *et al.* [21] show interesting changes in
the Raman spectra in $GdBa_2Cu_3O_{7-\delta}$ near 110°C and a phase change
at this temperature which could be ferroelectric. Recent studies of the
low-temperature specific heat in $YBa_2Cu_3O_{7-\delta}$ by Lawless *et al.*
(private communication) show features expected for a polar structure
and Scott [22] has also reported ferroelectricity below 240 K in that
crystal. Very recently Testardi [23] has made dielectric measurement
on the semiconducting tetragonal YBC and finds permittivity values in
the range 200 to 800, a clear indication of ferroelectric or antiferroelectric ordering.

It is most interesting to note the strong similarity of all the new high-T_c oxides to analogous ferroelectric structures, and we speculate that this similarity is more than accidental.

6. CONCLUSIONS

This chapter has focused upon past, present, and future possibilities for ferroic crystals, drawing mostly upon ferroelectric and ferroelastic examples. Materials from these families are already used to advantage in high-K capacitors, transducers, and pyroelectric sensors, and for electro-optic modulation and switching. Continuing development promises new and improved performance, particularly in thin polycrystal and single-crystal films and a range of new and unusual applications for 'smart' composites which incorporate both sensors and responders in active feedback control.

The new high-T_c superconductors are an interesting new 'wild card' in the ferroic pack, and it is interesting to speculate upon new applications which might make use of the known, ferroelastic and probably ferroelectric couplings in these solids.

REFERENCES

1. L. E. Cross and R. E. Newnham, In *Ceramics and Civilization*, Vol. III, ed. W. D. Kingery, The American Ceramic Society, Ohio, p. 289, 1987.
2. K. Aizo and T. Hirai, *Reports on Autumn 1969 Meeting of the Physical Society of Japan*, 437 (1969).
3. R. E. Newnham, *American Mineralogist*, **59**, 906 (1974).
4. J. Valasek, *Physical Review Bulletin* **24**(5), 560 (1924).
5. R. E. Newnham and L. E. Cross, *Materials Research Bulletin*, **9**, 927 (1974).
6. R. E. Newnham and L. E. Cross, *Materials Research Bulletin* **9**, 1021 (1974).
7. C. B. Sawyer, *Proceedings IRE* **19**, 2020 (1931).
8. H. Mueller, *Physical Review*, **47**(23), 175 (1935); **57**(9), 829 (1940); **58**(6), 565 (1940); **58**(9), 805 (1940).
9. A. F. Devonshire, *Philosophical Magazine* **40**, 1040 (1949).
10. W. Cochran, *Advanced Physics*, **9**, 387 (1960).
11. P. W. Anderson, Moscow Conference on Dielectrics, December 1960.
12. K. Aizu, *Journal of the Physical Society of Japan*, **23**, 794 (1967).
13. K. Aizu, *Journal of the Physical Society of Japan*, **32**, 1287, (1970).
14. L. A. Shuvalov, *Journal of the Physical Society of Japan (Suppl.)*, **28**, 38 (1970).

15. K. Aizu, *Journal of the Physical Society of Japan,* **27,** 387 (1969).
16. A. Amin, M. J. Haun, B. Badger, H. McKinstry and L. E. Cross, *Ferroelectrics,* **65,** 107 (1985).
17. M. J. Haun, E. Furman, S. J. Jang, H. A. McKinstry and L. E. Cross. *Journal of Applied Physics,* **62**(8), 3331 (1987).
18. R. E. Newnham, *Japan Journal of Applied Physics.* **24** (Suppl.) 24-2, 16 (1985).
19. W. A. Smith, *Composite Piezoelectric Materials for Medical Ultrasonic Imaging—A Review,* IEEE Catalogue No. 86 CH 2358-0.
20. B. Horowitz, G. Barsch and J. A. Krumhansl, *Physical Review Bulletin,* **36,** 8895 (1987).
21. S. K. Kurtz, L. E. Cross, N. Setter, D. Knight, A. Bhalla, W. W. Cao and W. N. Lawless, *Materials Letters,* **6,** 317 (1988).
22. M. S. Zhang, C. Qiuang, S. Dakun, R. Ji, Z. Quin, Y. Zheng and J. F. Scott, *Solid State Communications* **65,** 987 (1988).
23. L. R. Testardi, W. G. Moulton, H. Mathias and H. K. Ng, *Physical Review,* **37,** 2324 (1988).

6

Ceramics with Electrochemical Functions: The Role of Charged Interfaces

B. C. H. STEELE

Centre for Technical Ceramics, Imperial College, London, UK

ABSTRACT

Ionic transport across charged interfaces is first surveyed with oxygen sensors, high-energy batteries and ceramic electrochemical reactors (CERs) providing appropriate technological examples of electrochemical/Faradaic reactions. Particular emphasis is given to CERs because new monolithic configurations should provide solid oxide fuel cell (SOFC) systems with power densities greater than $1\,MW/m^3$ and fuel/electricity conversion values greater than 60%. Examples of the versatility of CER systems are provided by a selection of the reactions that can be carried out to produce useful chemical and electric power. The importance of charged interfaces in semiconducting ceramics is also illustrated by selecting relevant examples of positive-temperature-coefficient resistors, varistors, semiconducting gas sensors. Characterisation of both electrical and microstructural features in these electronic ceramics by electron beam-induced current (EBIC) and cathodoluminescence (CL) techniques is emphasised. Finally, preliminary remarks are made about the role of charged interfaces in the degradation of ceramics by stress corrosion.

1. INTRODUCTION

The processing of ceramic powders to form agglomerates of appropriate size, morphology, strength and toughness is dominated by the

103

charge on the surface of these particles. This solid–liquid interface is one example of a charged interface which provides the theme of the present contribution.

Charged interfaces will always be present in ceramic materials unless the concentration of charge carriers is very high, with the solid exhibiting metallic or possibly degenerate semiconductor behaviour. The development of the interfacial electrical potential is influenced by many factors during the fabrication process, including the presence of adsorbed species, segregation phenomena, furnace cooling rates, etc., which provide opportunities for the non-equilibrium distribution of charged species. Theoretical models for this complex situation in-

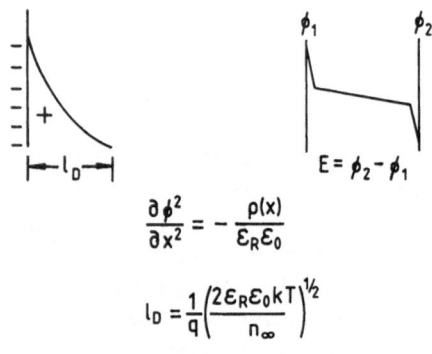

$$\frac{\partial \phi^2}{\partial x^2} = -\frac{\rho(x)}{\varepsilon_R \varepsilon_0}$$

$$l_D = \frac{1}{q}\left(\frac{2\varepsilon_R\varepsilon_0 kT}{n_\infty}\right)^{1/2}$$

l_D typically: 10 - 1000 Å

high field: 10^6 V/cm

rate constant: $k_r \propto \exp\left(-\frac{e\phi}{kT}\right)$

at 300 K, k_r changed by:
approx. 10 when ϕ changed by 0.1 V
approx. 10^9 when ϕ changed by 1 V

Lippman Equation :
relating surface energy (γ) and potential (ϕ) :

$$\left(\frac{\partial \gamma}{\partial \phi}\right)_T = -q_s$$

Note: surface energy decreases with
increasing surface charge (q_s)

FIG. 1. Summary of selected expressions for charged interfaces.

variably involve appropriate solutions of the Poisson–Boltzmann equation, which provide expressions for the surface potential (ϕ), Debye length (l_D), capacitance, and other parameters associated with the near-surface space-charge region. These expressions are summarised in Fig. 1, which also emphasises that very large electrical fields ($\sim 10^6$ V/cm) can be generated at these interfacial regions. In turn, these large fields can have major influences upon the rate processes at interfaces and upon the magnitude of the surface-energy value.

Experimental data on liquid–solid interfacial potentials can be obtained relatively easily by zeta potential and electrophoretic measurements, but a variety of techniques usually have to be employed to obtain information about gas–solid and solid–solid interfacial potentials. Direct measurements using vibrating-plate capacitance techniques are difficult, and it is usual to collect data by indirect a.c. and d.c. techniques. These can be supplemented by electron beam-induced current (EBIC) and cathodoluminesence (CL) investigations. Interpretation of the electrical measurements is assisted by chemical and structural investigations of the interfacial regions using TEM, SEM, SIMS, ESCA, AES, IR and Raman spectroscopy, surface EXAFS, etc.

The influence of charged interfaces upon the properties and application of a variety of ceramic materials is surveyed in the subsequent sections.

2. IONIC TRANSPORT AND CHARGED INTERFACES

2.1. Electrochemical/Faradaic Processes

2.1.1. Sensors
The widespread adoption of the potentiometric lambda probe for monitoring the oxygen content of automobile exhaust gases and boiler flue gases has been a major success for the advanced ceramics industry, and reviews are available [1] summarising the operation and performance of these devices. Whilst these sensors are suitable for distinguishing between oxidising and reducing conditions in burnt gases they are not appropriate for controlling the air–fuel ratio in 'lean-burn' car engines and pre-mixed domestic gas boilers. Accordingly a variety of amperometric oxygen monitors are being evaluated for these applications. These can be of the oxygen pump/gauge design

[2], or limiting current mode [3]. A limiting current assembly is depicted in Fig. 2a and typical performance data are shown in Fig. 2b.

Amperometric operation requires an improved specification for the ceramic electrolyte compared to potentiometric (zero-current) operation. High currents are required to obtain the diffusion-limited condition and so the electrical conductivity must be optimised with minimum grain-boundary resistance. At the same time the components should be thin ($\sim$200 μm) to allow high ionic fluxes and so the ceramic is required to be strong and tough. The complex impedance results (Fig. 3) for a series of tetragonal zirconia (TZP) ceramics indicates how the total specific resistivity can be decreased [4] by attention to powder purity and processing and TZP ceramics are now being evaluated for diffusion-limited devices.

2.1.2. High-energy batteries

By definition, the electrolyte phase in a high-energy battery is subjected to large electrochemical potential gradients. Very few ceramics at elevated temperatures are able to withstand both the reducing conditions imposed by the active negative-plate material (e.g. Li, Na) and the oxidising conditions associated with positive-plate components (e.g. S, Cl), and charging potentials in excess of 5 V invariably produce degradation of the ceramic electrolyte. This is not surprising when it is recalled that the electric fields associated with the electrochemical double-layer (space-charge) region can exceed 10^6 V/cm. Nasicon electrolyte ceramics, for example, in contact with liquid sodium and sulphur, rapidly degrade during the charging regime. In contrast, high-energy batteries incorporating beta" Al_2O_3 ceramics can withstand many thousands of cycles [5], providing inhomogeneous electric fields are not generated owing to the presence of cracks and second phases in the ceramic or interfacial polarisation associated with non-wetting of the active electrode components. Field-assisted crack growth (mode I failure) has often been discussed and is considered further in Section 4. Another type of failure mechanism in Na/S batteries can be attributed to electrolytic degradation (mode II failure). The molten sodium in contact with the beta-alumina ceramic electrolyte will slowly reduce the ceramic according to reactions of the type:

$$O_O^x \rightarrow [O]_{Na} + V_O^{\cdot\cdot} + 2e'$$

It is to be expected, therefore, that significant electronic conductivity

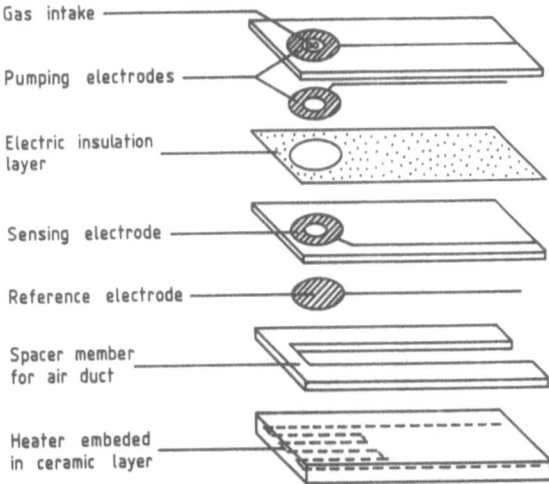

FIG. 2a. Components for amperometric zirconia oxygen monitor designed to operate in current-limiting mode.

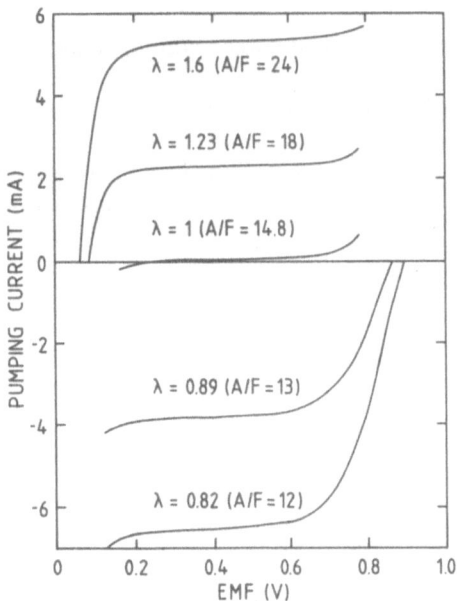

FIG. 2b. Performance characteristics of diffusion-(current-) limited oxygen monitor. A/F denotes air/fuel ratio.

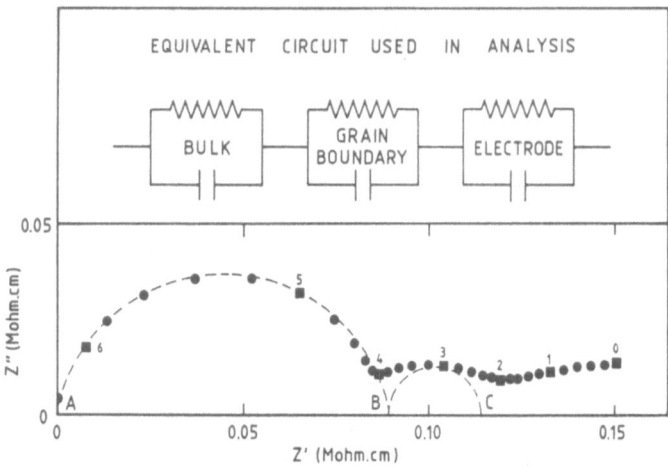

FIG. 3. Complex impedance data for tetragonal zirconia electrolyte. The magnitude of the grain-boundary impedance (second arc) can be influenced by preparative conditions and silica impurities.

could develop in the beta-alumina ceramic adjacent to the molten sodium, as depicted in Fig. 4 by the hatched zone. During the charging regime the electronic conductivity present can provide the necessary electrons to allow the Faradaic reaction to proceed within the solid electrolyte and the deposited sodium can be responsible for microcracking. Preliminary investigations by de Jonghe *et al.* [6] suggested

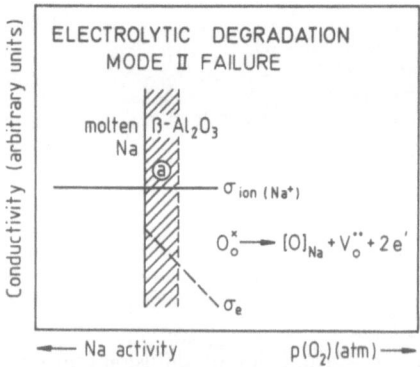

FIG. 4. Schematic diagram of development of mixed conductivity in beta-alumina ceramics.

that mixed conductivity in beta-alumina could appear relatively early in cycling tests. Accordingly, at Imperial College, McHale *et al.* [7] measured the oxygen self-diffusion in β-Al$_2$O$_3$ using ^{18}O/^{16}O exchange techniques in association with dynamic SIMS to establish the diffusion profile in the solid. Whilst these oxygen diffusion measurements tended to confirm the plausibility of the de Jonghe hypothesis, other observations have indicated that Na/S cells can perform satisfactorily for many thousands of cycles. It would appear, therefore, that the rate of the mode II degradation process is controlled by the magnitude of the electronic conductivity and that the investigations of de Jonghe were performed on poor-quality beta-alumina ceramics.

Finally, it should be noted that whilst the Na/S battery is still under active development in Europe and Japan, some of the groups are now evaluating the alternative system [8]

$$Na/\beta\text{-Al}_2O_3/FeCl_2(NaAlCl_4)$$

which produces much less corrosion of the container and current-collector materials.

2.1.3. Ceramic electrochemical reactors

There is a renewed interest world-wide in fuel cells for terrestial civil applications. In this area the phosphoric-acid fuel cell (PAFC) is close to commercialisation but is unlikely to achieve large-scale market penetration owing to its relatively low (~40%) chemical to electrical conversion efficiency (Fig. 5) and high capital cost. The high-temperature fuel cells, molten carbonate (MCFC), and solid oxide (SOFC), are now receiving increased attention because projections indicate much higher electrical generation efficiencies (55–60%). The high fuel/electricity conversion efficiency predicted for high-temperature fuel cell systems is associated with the in-situ reforming of hydrocarbon fuels, thus removing the requirement for an external fuel-processing stage which is an integral part of the PAFC system. Moreover, these high-temperature systems are particularly suitable for combined heat and power (CHP) applications.

Recent advances in SOFC technology now make this system an attractive option compared to MCFC designs. A major advantage of SOFC units is that the materials selected for the anode (Ni–ZrO$_2$ cermet), cathode (La$_{0.8}$Sr$_{0.2}$MnO$_3$), electrolyte (Zr(Y)O$_{2-x}$) and interconnector (LaCr$_{0.8}$Mg$_{0.2}$O$_3$) have been successfully tested over long periods (~30 000 h) without any significant degradation. Moreover, the technology of SOFC systems has benefitted from the recent

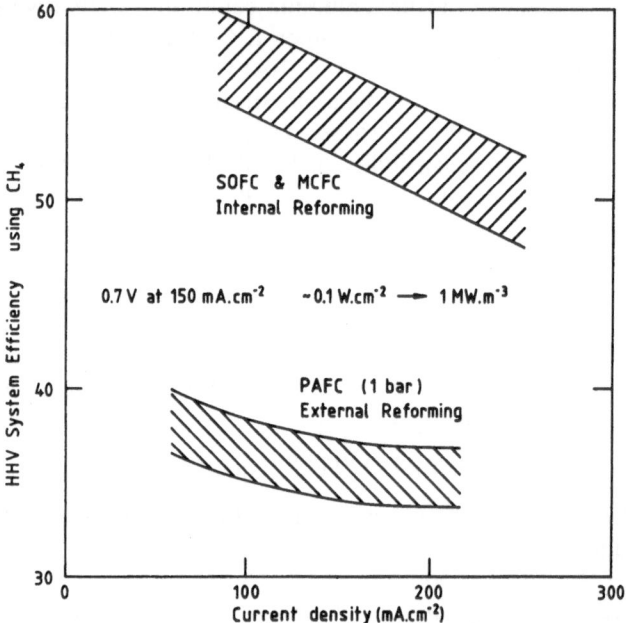

FIG. 5. Relationship between HHV (high heat value) fuel/electricity conversion efficiencies and current density for fuel cells operating at high (SOFC and MCFC) and low (PAFC) temperatures. (From J. Appleby, *Energy*, **11**, 21 (1986).

development of strong, tough, zirconia-based components for mechanical engineering applications. A variety of excellent powders are now available and the associated processing technology is well established.

For historical reasons the tubular configuration (Fig. 6) has received most attention and pilot-plant facilities operated by Westinghouse and Dornier are now available for the fabrication of units in the 1–5 kW range. However, alternative multichannel monolithic designs (Fig. 7) are now being evaluated because the high area/volume ratios associated with these monolithic units confer further benefits for the SOFC systems. It should be possible to fabricate such units with power densities approaching 1 MW/m³. Alternatively, these reactor designs could be operated at lower temperatures (800–900°C), which would result in lower power densities but higher fuel/electricity conversion values (>60%).

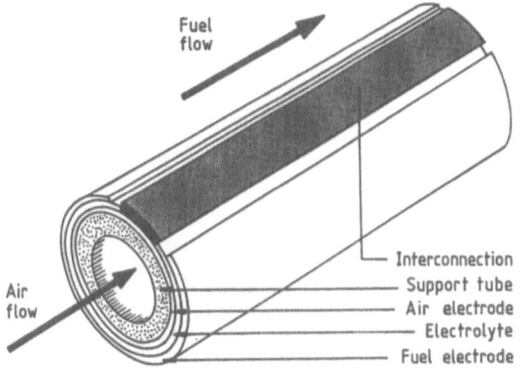

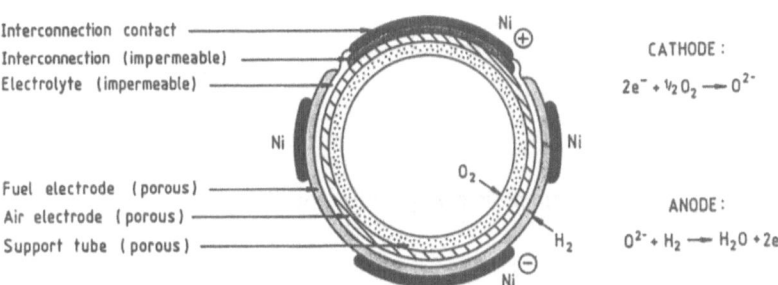

FIG. 6. Tubular configuration for SOFC system (Westinghouse design).

At present most performance data for SOFC systems has been obtained using syngas fuels (CO, H_2, H_2O) obtained from coal-gasification processes or steam reforming of hydrocarbons. At Imperial College we have recently demonstrated [9] that methane can be electrochemically oxidised directly to CO_2 and H_2O in small laboratory solid oxide cells. The standard anode ($Ni-ZrO_2$) has to be replaced by an oxide electrocatalyst, and by appropriate manipulation of the oxide anode composition it is possible to obtain satisfactory performance 0.7 V at 150 mA/cm^2 (0.1 W/cm^2) using 100% CH_4 feedstock. The advantages of oxide anodes compared to platinum anodes are clearly demonstrated in Figs 8 and 9. Examination of the relevant voltagram in Fig. 8 reveals that on the anodic sweep with platinum anodes the current only starts increasing at relatively anodic potentials (approximately $+0.1$ V). The potential at which this current increases is

B. C. H. Steele

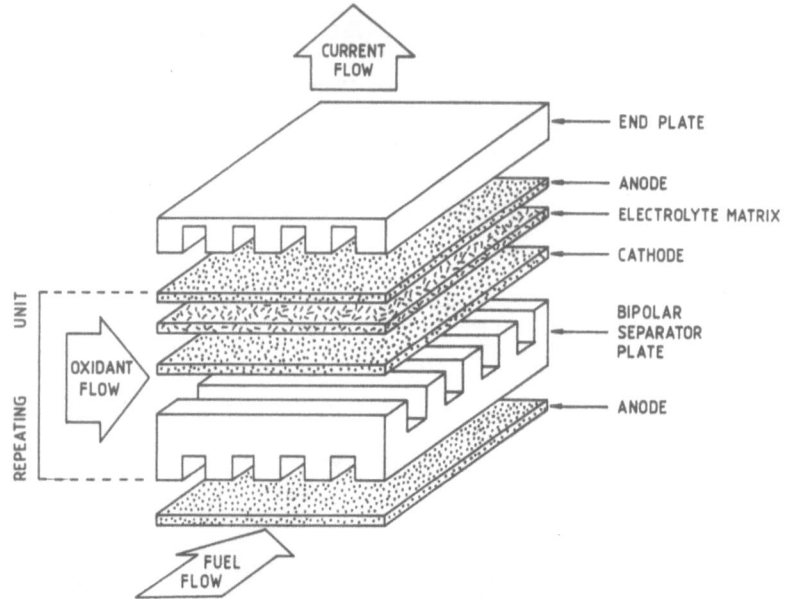

Fɪɢ. 7. Monolithic configuration for SOFC system.

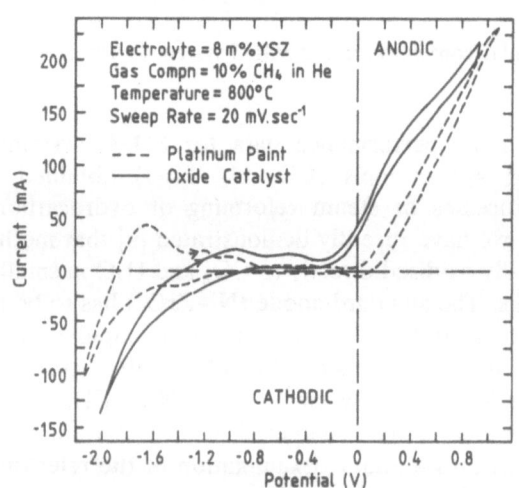

Fɪɢ. 8. Cyclic voltagrams for oxidation of methane using platinum and metal
oxide anode.

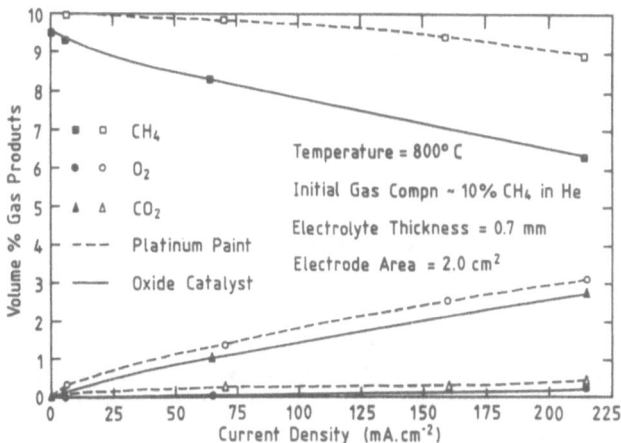

FIG. 9. Gas chromatographic analyses of anode product gases as a function of current density and anode material.

only changed very slightly (to $+0.15$ V) if pure helium is used instead of the CH_4/He mixture, and so it can be concluded that the increasing current is principally associated with the generation of gaseous oxygen molecules at the anode, which is confirmed by the gas-chromatographic results given in Fig. 9. However, with an appropriate oxide anode the curve in Fig. 8 indicates that the potential at which the anodic current commences to increase rapidly has been shifted about 0.4 V (from $+0.1$ to -0.3 V) to more cathodic values, which suggests that the oxide electrocatalyst is effective in promoting the oxidation of CH_4. This interpretation is confirmed by Fig. 9. As the current is increased, the product gas composition contains increasing amounts of CO_2 and decreasing amounts of CH_4, whilst the quantity of oxygen measured in the gas chromatograph remains very small even at current densities as high as 200 mA/cm^2. Improvements to the oxide electrocatalyst now allow useful current densities at -0.7 V to be realised as mentioned earlier.

Alternatively, by changing the anode electrocatalyst and operating the cell under potentiostatic control, it is possible to produce a variety of useful chemicals via partial oxidation reactions such as the oxidative coupling of methane to produce ethane [9]:

$$2CH_4 + \tfrac{1}{2}O_2 \rightarrow C_2H_6 + H_2O$$

Further, hydrogen can be produced by the high-temperature electrolysis of steam [10], and oxygen can be separated from air and other gaseous mixtures. The ability to operate in a variety of different modes is unique to ceramic electrochemical reactors, which means that investment in this general area of technology can be more readily justified as the development risk is spread over different market sectors.

It would be attractive to incorporate the strong, tough zirconia-based ceramics as components in the monolithic configurations. Research at Imperial College has demonstrated [11] that these materials can exhibit appropriate ionic conductivity values but little is known about the long-term stability of these ceramics in fuel-cell environments. It is possible that the metastable particles present in these tough ceramics will be induced to transform when exposed to steam at high temperatures (800–1000°C). Moreover, fuel-cell operation over extended periods involves a flux of oxygen ions in one direction. Accompanying this flux there is a tendency for cation migration in the opposite direction. This demixing process is negligible in cubic, fully stabilized zirconia but its magnitude and effects remain unexplored for the metastable toughened ceramics. Mechanical property behaviour under these conditions is just beginning to be investigated, as summarised in Section 4.

The performance of fuel cells is governed by a general relationship of the type

$$E = E^0 - IR - \eta_a - \eta_c$$

where E^0 is the open-circuit voltage of the cell, IR represents the ohmic losses, and η_a, η_c are the electrode activation and concentration polarisation losses, respectively. The concentration polarisation losses (η_c) can be minimised by appropriate design features to reduce hydrodynamic barriers to mass transport, but the activation overpotential (η_a) can be more difficult to manipulate. For example, with porous platinum cathodes on $Zr(Y)O_{2-x}$ electrolytes it has been established [12] that the rate-determining step at elevated temperatures (>450°C) in the reduction of oxygen is the diffusion of the charged O^{2-} surface species to an empty oxygen ion vacancy site, $V_O^{\cdot\cdot}$. The number and distribution of these vacancies is controlled by the surface segregation of yttrium, and their mobility is influenced by the formation of $[V_O^{\cdot\cdot} - Y_{Zr}']$ defect complexes. As yttrium segregates to the surface, the electrode polarisation increases. Confirmation that

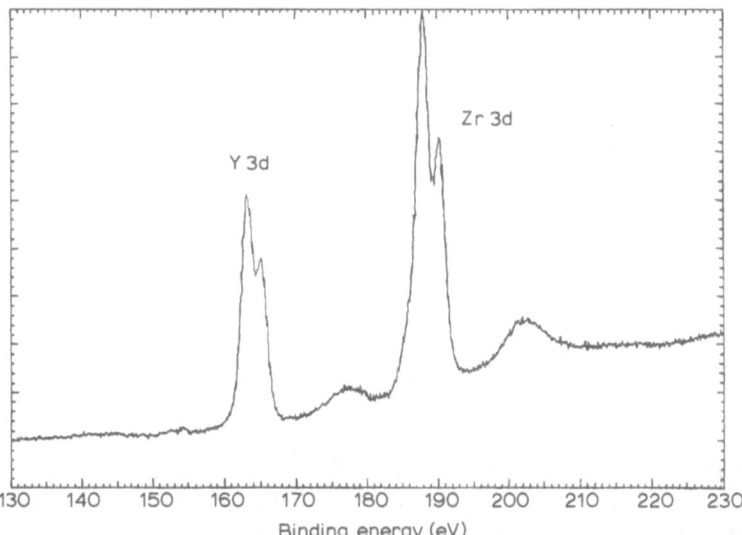

FIG. 10. Results of ESCA examination of (100) surface of 15% Y-doped ZrO_2 sample. Areas under peaks are proportional to concentration of each element at the surface.

segregation can occur is provided by ESCA examination of the surfaces of single crystals of $Zr(Y)O_{2-x}$ having the bulk composition $Zr_{0.85}Y_{0.15}O_{1.925}$. The results are reproduced in Fig. 10. The areas under the peaks are proportional to the relative concentrations of Zr^{4+} and Y^{3+} at the surface, and it can be seen that the ratio Y/Zr is approximately 1:2 at the surface compared to the bulk value of 1:5·7. That segregation of the yttrium ion has occurred is confirmed by removing the surface layer by ion-beam milling, and subsequently re-analysing the sample. After this procedure, the Y^{3+} peak intensity is reduced so that the Y/Zr ratio is around 1:6.

Finally, it should be noted that for CHP applications it would be desirable to operate SOFC units around 700–750°C. Improved oxygen ion conductors would be required for this temperature of operation. The possibilities of obtaining such ceramic electrolytes which satisfy the additional specifications for fuel-cell components are discussed by Steele in a recent review [13] and it is interesting to record that $BaCe_{0.85}Gd_{0.15}O_{2.925}$ satisfies at least the anionic conductivity requirements (N. Bonanos and M. Mahmood, private communication).

2.2. Composite Electrolytes

Enhanced ionic conductivity in multiphase systems has now been demonstrated for many systems since the early reports by Liang [14] for LiI–Al$_2$O$_3$. Although the details of quantitative interpretations often remains controversial, the overall qualitative features are clear. An interfacial space-charge is established which can either enhance or deplete the concentration of ionic charge carriers in the boundary region, as depicted in Fig. 11 for a cation Frenkel-disordered solid such as LiI or AgI. The properties of this interfacial region have been examined in detail by Maier [15], and its influence on the total bulk conductivity is demonstrated in Figs 2(a) and (b). When the concentration of the dispersed phase (e.g. Al$_2$O$_3$) is small, most of the particles are completely surrounded by the matrix conductor phase (Fig. 12(a)) and so there is not a continuous enhanced conduction path through the solid. As the concentration of the dispersed phase is increased, the interfacial regions become linked together to produce a maximum in the total conductivity. Increasing the volume fraction of the second phase still further results in a decrease in the total conductivity, as the number of dispersed phase particles which are not completely covered by the interfacial boundary layer will become smaller. An increasing volume concentration of the poorly conducting phase will also produce a decrease in total conductivity, as the fraction of bulk percolation pathways will fall. Thus, the variation of total conductivity will depend upon the volume fraction of the second phase and the particle size of that dispersed phase, which has often been

ENHANCED CONDUCTIVITY IN MULTIPHASE SYSTEMS

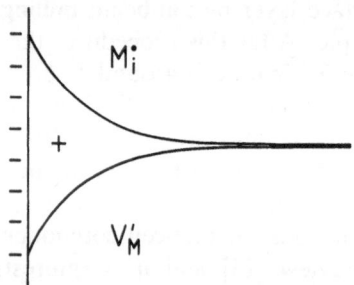

Fig. 11. Schematic profile of point defects in the interfacial region of a Frenkel-disordered solid (e.g. LiI. AgI).

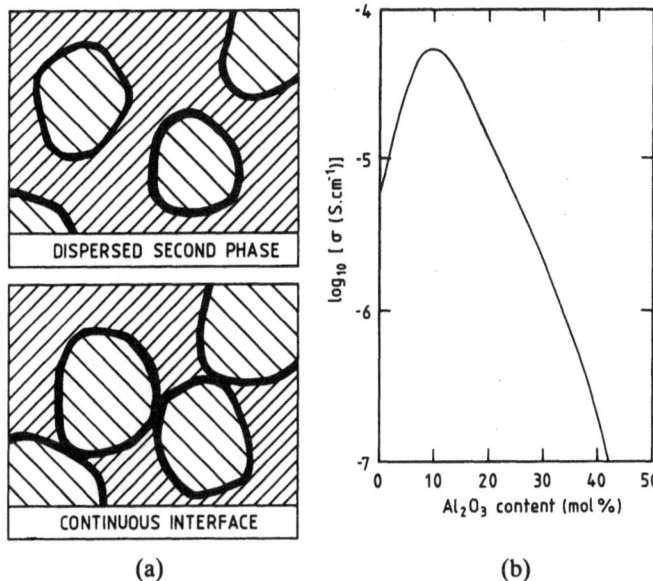

(a)   (b)

FIG. 12. (a) Arrangement of interfacial enhanced conductivity region as a function of concentration of dispersed second phase. (b) Ionic conductivity as a function of concentration of dispersed second phase.

confirmed for cation Frenkel-disorded solids. More recently, enhanced conductivity has been reported for anion Frenkel-disordered solids and results for $CaF_2-Al_2O_3$ reported by Fujitsu *et al.* [16] are shown in Fig. 12. At Imperial College we have been examining the tetragonal ZrO_2-CeO_2 system with Al_2O_3 additions and again our preliminary investigations indicate a conductivity enhancement with a maximum occurring around 7 wt% Al_2O_3 (D. Gray, private communication).

3. SPACE-CHARGE PHENOMENA INVOLVING ELECTRONIC TRANSPORT

The electrical behaviour of grain boundaries in semiconducting ceramics is exploited in positive-temperature-coefficient resistors (PTCR), varistors, gas sensors, and electroluminescent display devices. The general properties of these technologically important materials have often been reviewed [17], but again details of the interfacial phenomena are still being investigated. Segregation of electrically active

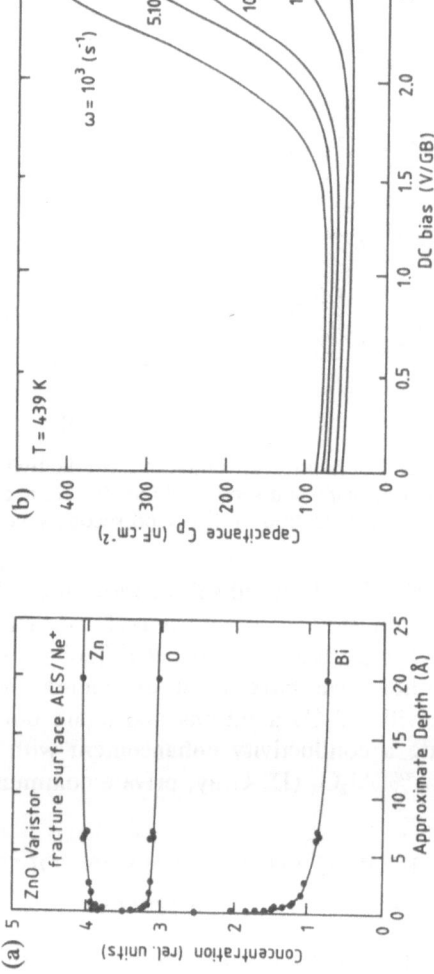

Fig. 13. (a) Concentration profiles of Bi, Zn and O obtained by Auger spectroscopy at a ZnO varistor grain boundary. (b) Capacitance (C_P) versus voltage V at several frequencies illustrating role of interfacial electronic states.

species, behaviour of the Schottky barriers, including the role of electronic traps, are being correlated with microstructural features. An example [18] of segregation effects in ZnO varistor ceramics is reproduced in Fig. 13(a), and the variation of the interfacial capacitance with potential and frequency shown in Fig. 13(b) provides data relating to the associated interfacial electronic states. These results allow a model interfacial energy band diagram to be constructed which can then be evaluated by a further series of experiments. In this context an important technique which produces both electrical and microstructural information is the application of electron beam-induced current (EBIC) and cathodoluminescence (CL) measurements in the scanning electron microscope (SEM). The experimental arrangement for EBIC and CL measurements are depicted in Fig. 14 and relevant theoretical aspects have been described in the literature [19]. The scanning electron beam generates electron–hole pairs, and those sufficiently close to the electrical field existing at space-charge junctions will be separated. These separated charges will produce a current which can be measured in an external circuit. The spatial distribution of the current can be displayed on a screen to show the location of the potential barriers at the grain-boundary regions. Further information about the potential barriers is provided by operating in the CL mode. The radiative recombination of electrons

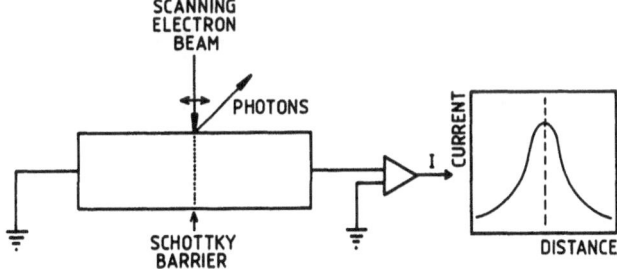

ELECTRON BEAM INDUCED CURRENT (EBIC):
image of current distribution

CATHODOLUMINESCENCE (CL):
characteristics of photons emitted due to
radiative recombination

Fig. 14. Schematic representation of experimental arrangements for EBIC and cathodoluminescence techniques.

and holes at traps in the interfacial regions produces photons which can be analysed in a spectrometer. In this manner the energetics of the traps can be determined and assigned to specific defects such as barium vacancies in $BaTiO_3$-based PTCR devices. A useful example of the application of this technique is provided by the investigations of Koschek and Kubalek [20].

Modification of the surface potential by adsorbed species is often used to manipulate the behaviour of liquid–powder suspensions. In another area, the ability of adsorbed species to modulate the electrical properties of semiconductor oxides is exploited in the Taguchi type of gas sensor, which is usually based on *n*-type doped SnO_2 ceramics [21]. In air, oxygen is adsorbed at interfacial boundaries to produce

(a)

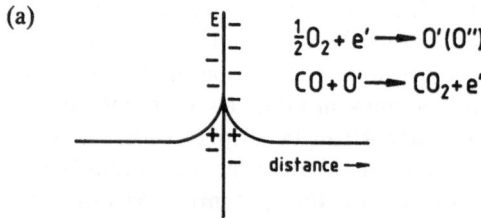

Doped SnO_2 as a sensor for reducing gases.

(b)

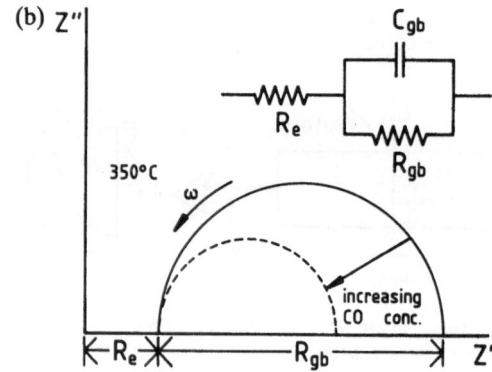

FIG. 15. (a) Schematic representation of development of charged interfacial regions in SnO_2 sensors due to adsorption of oxygen molecules at S. (b) Complex impedance data for SnO_2 sensor showing influence of CO upon electrical behaviour of interfacial regions.

negatively charged surface species and adjoining high-resistance depletion space-charge regions as depicted in Fig. 15a. In the presence of reducing gases such as CO, the concentration of adsorbed oxygen is reduced by reactions such as,

$$O_s' + CO \rightarrow CO_2 \uparrow + e'$$

The electrons are re-injected into the space-charge region, producing a sharp decrease in resistance which can be monitored by an appropriate electronic circuit. Changes in the interfacial barrier can also easily be investigated by a.c. impedance spectroscopy, as shown in Fig. 15b, which summarises results obtained at Imperial College.

4. DEGRADATION PROCESSES

It has often been recognised [22] that surface potentials are a significant parameter in the stress corrosion of oxide and silicate materials. Moreover, the surface charge can be influenced by the presence of adsorbed species such as OH^-. However, there does not appear to be much systematic data for ceramic systems relating surface potentials to degradation processes in aqueous systems. At somewhat higher temperatures ($\sim 300°C$), crack propagation in β-Al_2O_3 ceramic electrolytes was extensively studied about 10 years ago. Models were developed for this degradation process, which was often termed the mode I failure mechanism. The general assumption in these models was that during the charging cycle sodium was deposited at the crack tip and that a Poiseuille pressure was generated which provided the stress for crack propagation. However, the calculated critical current densities [23, 24] were orders of magnitude greater than the experimentally measured values, and so special situations were often invoked requiring very high localised currents. However, it is also probable that very high fields ($\sim 10^5$ V/cm) are developed at crack tips. This could lead to the injection of electrons (cf. mode II failure mechanism mentioned in Section 2.1.2) and subsequent deposition of sodium in the solid electrolyte ahead of the crack tip. High fields would also produce a significant decrease of the effective surface energy via the Lippman equation,

$$\left(\frac{\partial \gamma}{\partial \phi} \right)_T = q_s$$

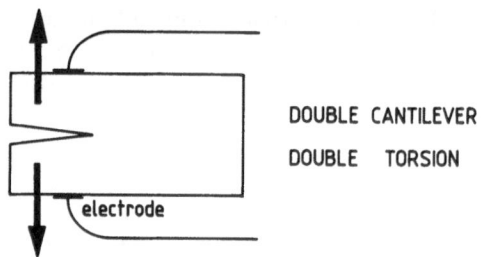

FIG. 16. Schematic diagram showing arrangement for investigating slow crack growth using a.c. impedance spectroscopy with d.c. bias.

where γ, ϕ, q_s represent the surface energy, potential and charge, respectively. This would imply that the local critical stress intensity factor, K_{IC}, would be much lower than values derived from bulk mechanical test procedures, which could provide an explanation for crack propagation at relatively low current densities.

The influence of interfacial space-charge regions and of ionic fluxes on the mechanical behaviour of zirconia-based ceramic electrolytes is of major concern for the operation of ceramic electrochemical reactors (see Section 2.1.3). Accordingly, experiments are under way at Imperial College to investigate the role of these parameters on the rate of crack growth in toughened zirconia ceramic electrolytes. The extension of crack notches in flat plates of ceramic electrolytes using double-beam cantilever or double-torsion techniques is being monitored by a.c. impedance measurements as depicted in Fig. 16. The use of the electrical potential drop method for measuring crack extension rates in ceramics has been analysed by Nicholson *et al.* [25]. More information can be obtained from impedance spectrocopic measurements, as changes in capacitance can be monitored and d.c. biases may also be imposed to produce ionic fluxes transverse to the crack propagation direction. Preliminary experiments indicate that the application of a d.c. bias can have significant effects on the rate of slow crack growth.

Finally, it should be noted that the presence of electrically active defects on fracture surfaces can also be detected by EBIC and CL techniques and reports are available in the literature [26].

5. CONCLUSIONS

Ceramics with electrochemical functions already provide one of the more technologically successful applications of advanced ceramics in the form of lambda sensors, and ceramic electrochemical reactors are expected to be widely exploited in the next decade. These solid-state electrochemical systems involve mass transport and Faradaic reactions at charged interfaces. The electrical behaviour of grain boundaries in semiconducting ceramics is also exploited in a wide variety of devices and many of the techniques used to characterise solid-state electro-chemical interfaces can be used to examine electronic transport across charged interfaces and vice versa. It is also anticipated that the behaviour of charged interfaces will be recognised in the future as having a much greater role than hitherto in degradation and crack propagation processes.

ACKNOWLEDEMENTS

The author wishes to thank past and present members of the Centre for Technical Ceramics at Imperial College for their invaluable discussions and contributions. The participation of Dr D. B. Holt in collaborative work on the application of EBIC and CL techniques to electrical ceramics is particularly acknowledged.

REFERENCES

1. W. C. Maskell and B. C. H. Steele, Solid state potentiometric oxygen gas sensors. *J. Appl. Electrochem.*, **16**, 475 (1986).
2. D. M. Haaland, *Analyt. Chem.*, **49**, 1813 (1977).
3. K. Saji, Characteristics of limiting current-type oxygen sensor. *J. Electrochem. Soc.*, **134**, 2430 (1987).
4. C. A. Leach, P. Tanev and B. C. H. Steele, Effect of rapid cooling on the grain boundary conductivity of Yttria partially stabilized zirconia. *J. Mat. Sci. Lett.*, **5**, 893 (1986).
5. J. Sudworth and A. R. Tilley (eds), *The Sodium Sulphur Battery*. Chapman & Hall, London, 1985.
6. L. C. de Jonghe, L. Feldman and A. Beuchele, Slow degradation and electron conduction in sodium/beta aluminac. *J. Electrochem. Soc.*, **16**, 780 (1981).
7. A. E. McHale, J. A. Kilner and B. C. H. Steele, Determination of oxygen diffusivities in β and β'' alumina by $^{18}O/^{16}O$ exchange. In *Transport in Nonstoichiometric Compounds* ed. G. Simkovich and V. S. Stubican. Plenum Press, New York, 1985, p. 217.
8. J. Coetzer, *J. Power Sources*, **18**, 377 (1986).
9. B. C. H. Steele, I. Kelly, H. Middleton and R. Rudkin, Oxidation of methane in solid state electrochemical reactors. *Solid-State Ionics* **28/30**, 1547 (1988).

10. W. Donitz, G. Dietrich, E. Erdle and R. Streicher, In *Proc. 6th World Hydrogen Conf., Vienna* ed. T. N. Vezirohlu, N. Getoff and P. Weinzierl, 1986, p. 271.

11. N. Bonanos, R. K. Slotwinski, B. C. H. Steele and E. P. Butler, High ionic conductivity in polycrystalline tetragonal Y_2O_3–ZrO_2. *J. Mater. Sci. Lett.*, **3**, 245 (1984).

12. J. Mizusaki, K. Amano, S. Yamauchi and K. Fuecki, Electrode reaction at Pt, $O_2(g)$/stabilized zirconia interfaces. *Solid-State Ionics* **22**, 313 (1987).

13. B. C. H. Steele, Oxygen ion conductors. In *High Conductivity Solid Ionic Conductors,* ed. T. Takahashi, World Scientific, Publ., Singapore, 1989.

14. C. C. Liang, Conduction characteristics of the lithium iodide-aluminium oxide solid electrolytes. *J. Electrochem. Soc.,* **120**, 1289 (1973).

15. J. Maier, Defect chemistry and conductivity effects in heterogeneous solid electrolytes. *J. Electrochem. Soc.,* **134**, 1524 (1987).

16. S. Fujitsu, M. Miyayama, K. Koumoto, H. Yanagida and T. Kanazawa, Enhancement of ionic conduction in CaF_2 and BaF_2 by dispersion of Al_2O_3. *J. Mater. Sci.,* **20**, 2103 (1985).

17. L. M. Levinson (ed.), *Grain Boundary Phenomena in Electronic Ceramics.* Advances in Ceramics, **1** (1981).

18. M. Rossinelli, G. Blatter and F. Greuter, *Br. Ceram. Proc.,* **36**, 1 (1985).

19. D. B. Holt and M. Lesniak, *Scanning Electron Microscopy,* **1**, 67 (1985).

20. G. Koschek and E. Kubalek, Grain-boundary characteristics and their influence on the electrical resistance of barium titanate ceramics. *J. Am. Ceram. Soc.,* **68**, 582 (1985).

21. J. F. McAleer, P. T. Moseley, B. C. Tofield and D. E. Williams, *Br. Ceram. Proc.,* **36**, 89 (1985).

22. T. A. Michulske and B. C. Bunker, Stress corrosion of mixed ionic/covalent solids. *J. Am. Ceram. Soc.,* **69**, 721 (1986).

23. L. A. Feldman and L. C. de Jonghe, Initiation of mode I degradation in sodium-beta alumina ceramics. *J. Mater. Sci.,* **17**, 517 (1982).

24. L. Viswanathan and A. V. Virkar, *J. Am. Ceram. Soc.,* **66**, 159 (1983).

25. T. B. Troczynski and P. S. Nicholson, Sensitivity of the potential drop technique for crack length measurement in chevron-notched specimen. *J. Am. Ceram. Soc.,* **69**, C-136 (1986).

26. T. F. Page and J. T. Czernuszka, Cathodoluminescence: A microstructural technique for exploring phase distributions and deformation structures in zirconia ceramics. *J. Am. Ceram. Soc.,* **68**, C-196 (1985).

7

Sol–Gel Processing of Optical Waveguides

JOHN B. MACCHESNEY

AT & T Bell Laboratories, Murray Hill, New Jersey, USA

ABSTRACT

This paper was written for inclusion in the program of the 'International Symposium on Advanced Ceramics' to be held under the auspice of the Tokyo Institute of Technology. It sketches the history of optical waveguide processing over the past two decades. Sol–gel is emphasized as the means for generating the next (fourth-) generation process. Examples are presented to demonstrate that fibers produced this way exhibit satisfactory optical and mechanical performance. Prospects for continued development to an eventual technology are discussed.

1. INTRODUCTION

The past decade has seen the installation of millions of kilometers of optical fibers in long-distance telecommunications networks. In the coming decade this transmission medium will extend toward the loop and subscriber network. Billions of kilometers of fiber will eventually be used for this purpose. One questions whether vapor deposition processes, so successful for long-haul fiber, will remain the process of choice in the coming era; can this be replaced by a forming process, such as sol–gel, which operates to provide bulk fabrication at room temperature?

Before discussing this possibility, let me put the technology of fiber fabrication into perspective. First attempts at making fiber occurred in

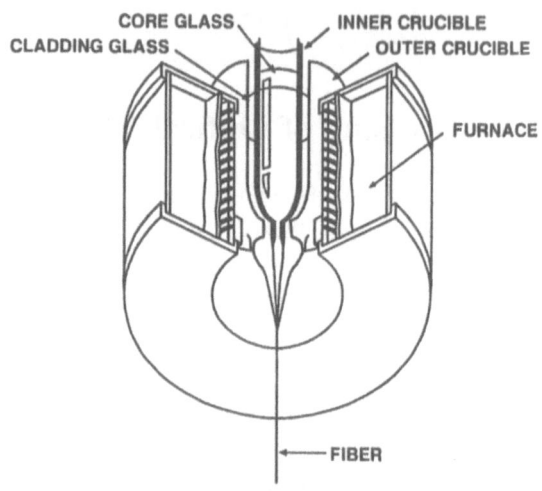

FIG. 1. Double-crucible process for optical fiber.

the mid-1960s. These used multicomponent glasses: soda–lime–silica, sodium borosilicate, and attempted simultaneous drawing of core and cladding from double crucibles [1] (one containing a core composition, the other cladding). This is shown in Fig. 1.

Just as the double crucible started to achieve acceptable properties for graded-index fibers, the vapor deposition processes appeared. In rapid sequence, outside vapor deposition (OVD) [2], modified chemical vapor deposition (MCVD) [3], and vertical axal deposition (VAD) [4], produced fibers whose properties approached the intrinsic limits of their principal constituent—silica. These are portrayed in Fig. 2. OVD and VAD deposit particles of silica ('soot'). In the former, a core composition (GeO_2–SiO_2) is first deposited on a mandrel followed by silica cladding. In the case of graded-index fibers, by varying the composition of the soot an approximately parabolic profile can be produced to minimize mode-disperion. VAD deposits end-on, the grading of the GeO_2-containing core occurs by virtue of a temperature gradient imposed across the deposition surface. The resulting soot blank of either is consolidated in He–Cl_2 atmosphere to produce a pore-free, low-OH^- silica glass. MCVD, on the other hand, simultaneously deposits and consolidates layers of soot deposited inside a silica substrate tube to yield core and cladding as required. The core composition can be changed from layer to layer (pass to

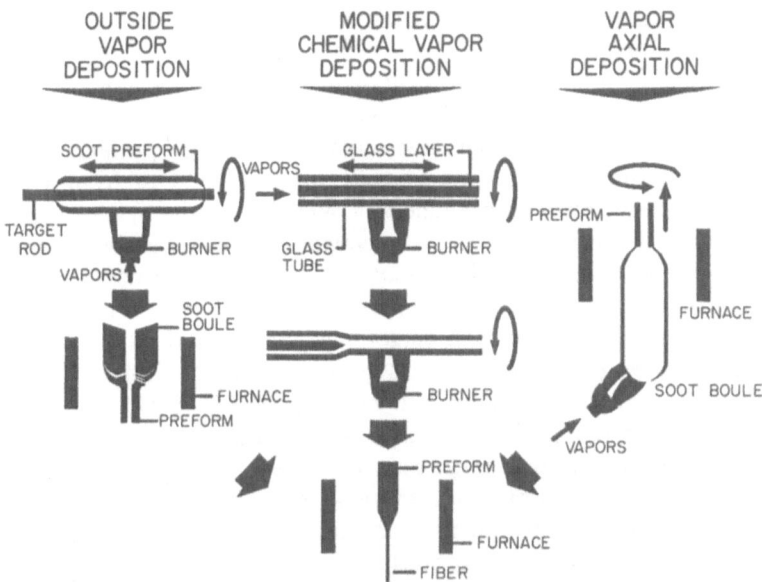

OUTSIDE VAPOR DEPOSITION

MODIFIED CHEMICAL VAPOR DEPOSITION

VAPOR AXIAL DEPOSITION

FIG. 2. Vapor-deposition processes OVD, MCVD, VAD.

pass of the torch traverse along the tube) to produce the graded index. After deposition, the tube is collapsed forming a preform rod.

The third generation of fiber processes are the response to changes in transmission systems design. In particular, multimode fiber, preferred for the early installers, has largely been replaced by single-mode. This fiber gives yet lower dispersion (by eliminating mode dispersion) because its small core supports the propagation of only the fundamental mode. Because of the small core diameter (typically only about 1/15 that of multimode fiber), overcladding can be employed. A consolidated (vitreous) preform made by any method can be overclad with soot or a vitreous silica tube to produce larger blanks, and these drawn to longer lengths of fiber. The overcladding material need not be of the same optical quality as the preform because in a single-mode fiber, although the evanescent field of light spreads beyond the core, power becomes negligible beyond approximately six core diameters. Thus, the quality of the silica overcladding beyond this region is of little consequence to the optical properties of the eventual fiber. Overcladding makes it possible to increase the size of preforms from a few tens of kilometers of fiber for the stand-alone processes to 100 or more kilometers.

Thus, the scene is set for the emergence of a new technology, one using inexpensive, commercially available starting material, not specially purified to remove transition metal ions to the ppb level as in vapor deposition. By this new process, tubes can be cast from a gel and used to overclad a core/cladding member made by the vapor deposition. Of course, how close the single-mode core can be brought to the gel-silica overcladding without diminishing the transmission of the light signal depends upon the quality of the gel-silica. Further, an eventual technology looks at an all-gel fiber, one comprised of a doped gel core and gel cladding as well as gel overcladding. Both strategies overcladding and all-gel fiber, are being pursued. However, to be successful, both require that large bodies (<1 kg mass) be dried and consolidated without breaking. This has been, and is, the greatest impediment to the evolution of a sol–gel technology.

Various approaches to the problem have been tried, but the basic sol–gel process is the same. It starts by the formation of a silica sol. This is gelled to form a tube which is air-dried. This tube is then fired in a manner reminiscent of that used to consolidate soot formed by OVD or VAD. Generally, firing consists of two stages: a purification stage at 800–1200°C in a chlorine-containing atmosphere and a consolidation step at 1350–1450°C. Figure 3 diagrams the sequence of steps.

A number of starting materials can be used to make the silica sol: tetraethyl orthosilicate ($Si[OC_2H_5]_4$), $SiCl_4$ or commercial fumed silica

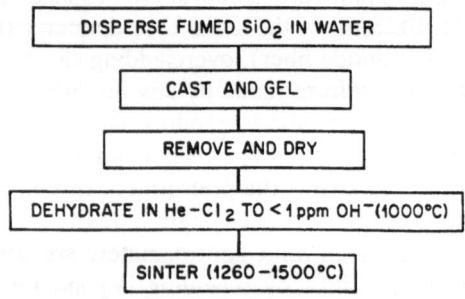

Fig. 3. Flow diagram for preparation of optical fiber preform.

POLYMERIZED GEL

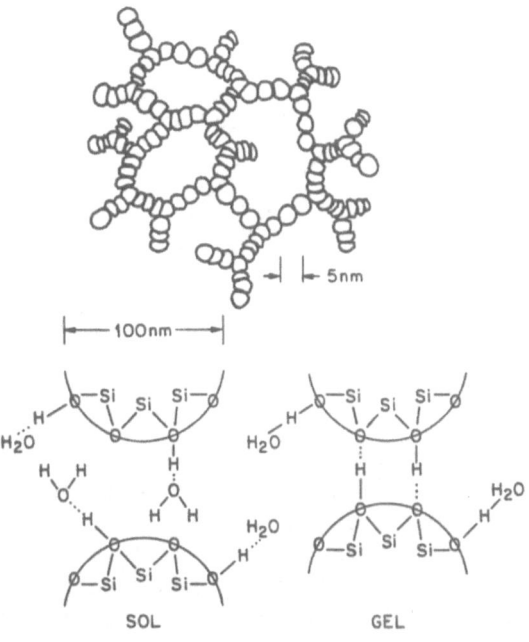

COLLOIDAL SOL AND GEL

FIG. 4. Schematic picture of polymeric silica gel compared to that of colloidal silica.

such as Cab-O-Sil† or Aerosil‡. The gels resulting from these are quite different. In the case of the first starting material, a polymerized gel results. This is characterized by small particle size and small pore size. Use of fumed silica allows a choice of particle sizes, and larger sizes—in the range of a fraction of a micrometer—are available. Colloids from this source are hydrogen-bonded to form the gel rather than polymerized siloxane linkages as in the former case. A representation of the two extreme cases are shown in Fig. 4.

The importance of starting particle size lies with the pore structure imparted to the wet gel. As mentioned previously, the greatest

† A product of the Cabot Co., Billrica, Massachusetts, U.S.A.
‡ A product of Degassa, A.G., Hanau, F.R.G.

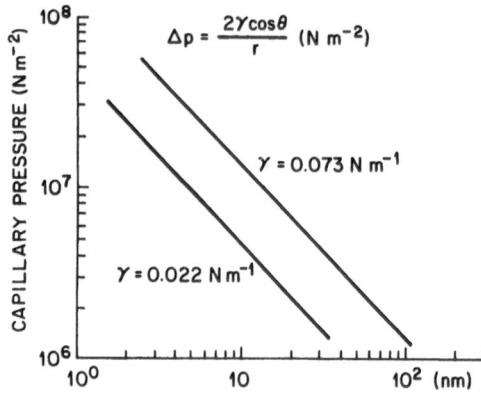

FIG. 5. Capillary pressure vs pore diameter for water and alcohol.

impediment to a sol–gel technology for producing large silica bodies is cracking and fracture during drying of the wet gel. This is influenced by capillary pressure developed as water-filled pores empty by evaporation from the surface. As illustrated by Fig. 5, which plots capillary pressure vs pore size for water and alcohol [5], it is apparent that both pore size and surface tension of the liquid significantly change the capillary pressure. Fractures caused by these capillary forces can be reduced or eliminated by a number of processing steps. For instance, the wet gels can be strengthened by addition of polymers or, otherwise, making the colloidal surface hydrophobic. Finally, such fractures can be totally eliminated by removing the liquid–vapor interface, as occurs by drying under supercritical conditions.

The supercritical approach has been successfully applied to drying of high-surface-area (small-pore-size) polymeric type gels made by hydrolysis of alkoxides [6]. In one instance an autoclave is used to dry the material. It is first pressurized with nitrogen, then heated to above critical conditions before the pressure is slowly bled off. This eliminates the liquid phase comprising more than 50% of the original gel without creating a liquid–vapor interface and the resulting capillary stresses.

In order to avoid the drying problem completely, colloidal silica can be compressed mechanically, without going through the gel route. This process developed by Dorn *et al.* [7], predensifies colloidal silica whose specific surface area ranges from 50–350 m²/g. This powder is then compacted in cylindrical molds to form a body where porosity runs from 71–90%. Since no water need be removed, stress cracking is not a problem. The resulting open porosity aids in the purification of the material during consolidation.

Certainly, purification of the rather impure commercial fumed silica is essential to the achievement of state-of-the-art fiber. The porous nature of silica bodies produced by these methods provides an opportunity to do this and results in just the inverse of the situation encountered in the earlier double-crucible process. You will recall that this failed because starting materials with impurity levels in the parts per billion range yielded glass with transition metal concentrations in the parts per million, the glass being contaminated during processing. The porous gel-derived silica is consolidated in a non-contaminating silica mantle and in a chlorine-containing atmosphere where significant removal of impurities and hydroxyl contamination occurs [8]. Here, both the presence of chlorine and the use of a low oxygen partial pressure atmosphere are important. This is understood from the reaction:

$$Fe_2O_3 + 2Cl_2 = 2FeCl_2 + \tfrac{3}{2}O_2$$

By firing in an atmosphere protected from air intrusion, oxygen partial pressures are in the range of 10^{-6} atm. Thus, at temperatures between approximately 800 and 1200°C, iron and other impurities are effectively removed. It was demonstrated that a gel intentionally contaminated with 1 wt% hematite contained only 40 ppb of iron after dehydration/consolidation [8].

Earlier, I outlined the general sol–gel procedure. Now I will give examples of two routes being pursued: for making the entire fiber from sol–gel, and a hybrid strategy wherein the core region (the optically active region) is obtained from vapor deposition while the vast bulk of the fiber is composed of gel-derived material. In the former, fibers are composed of a core gel made from $Si(OC_2H_5)_4$ and $Ge(OC_4H_9)_4$ [9]. However, the difficulty was encountered in preventing the crystallization of GeO_2 and its removal by solution due to the amphoteric nature of the germanium ion. A more successful approach [10] consisted of making a fluorine-doped gel tube using tetraethyl orthosilicate and the fluorine-containing analog $[Si(OC_2H_5)_3F]$. After drying, this is consolidated in He, Cl_2 and O_2. The now vitreous silica tube is heated to a yet higher temperature in oxygen to out-diffuse fluorine from a central shell of material and produce a pure silica core grading to a F-doped cladding with index difference $(\Delta n) = 0.3\%$.

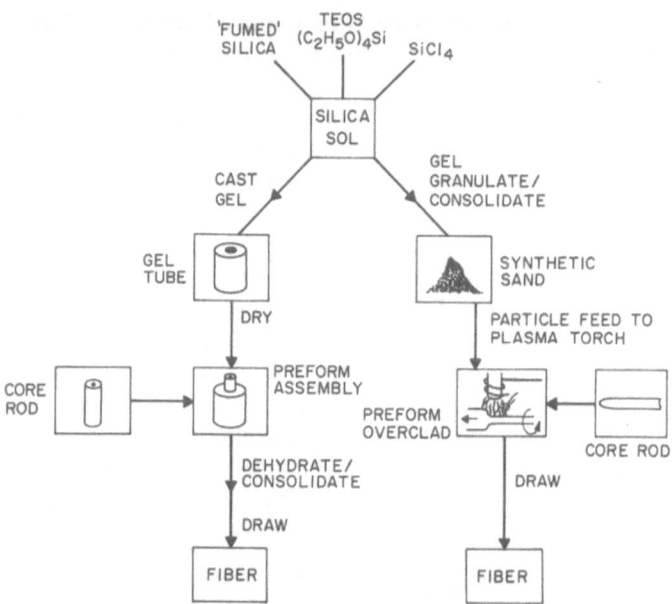

FIG. 6. Hybrid strategy.

Two approaches to the hybrid strategy are shown in Fig. 6. Starting from a variety of precursor materials, overcladding is accomplished by fusion of gel particles [11] or alternately by consolidation of colloidal gel [12]. Using either, a small quantity of conventionally (vapor-phase) prepared material is made to comprise the optically active portion of the fiber, while the bulk of the fiber (up to 95%) can be gel-derived.

Before describing these, let us digress for a moment and consider how silica is made today. Because it is refractory and its viscosity remains high even above 2000°C, it is not made by conventional melting techniques. Instead, particles of ground quartz, about $100 \, \mu$m in size, are melted in an oxy-hydrogen flame or RF plasma and are deposited in a thin layer on the face of a boule. Currently, naturally occurring silicas are used but powder of a size between 100 and $500 \, \mu$m can be prepared by granulating a gel [13]. Of course, this same technique can be used for overcladding. Particles, melted in a plasma [13] or flame [14] have the advantage over 'soot' overcladding that these particles have sufficient mass to be directed at a core member. Soot particles are 1000 times smaller, cannot be targeted and depend on weak thermophoretic forces to deposit. Plasma fusion of gel-derived particles has achieved rates of 15 g/min, which is generally higher than that obtained by soot deposition in spite of more extensive development efforts expanded in the latter.

The second approach overclads a core rod consisting of an up-doped core and some minimum thickness of vapor-deposited cladding. The essential feature here is that the evanescent field of the light transmitted by a single-mode fiber extends beyond the boundary of the core. Thus, some thickness of very low-loss cladding (as provided by vapor deposition) and a good core–cladding interface are essential to preparing state-of-the-art fiber by this method. Figure 7 shows the profile of a core rod used to produce fiber by these means. It can be shown that the case of a core rod with D/d of 4 (O.D. of the rod/diameter of the core) the contribution of a gel-derived overcladding (assuming its loss is 10 dB/km) to the total loss of the fiber would be only 0·001 dB/km.

Actually gel-derived silica is not this bad. Losses below 5 dB/km at 1·55 μm have been observed (Johnson, D. W., MacChesney, J. B. and Fleming, D. A., unpublished) for fibers consisting of a gel-silica (Cab-O-Sil, 90 m²/g) core overclad by down-doped MCVD. Losses below 1 dB/km have been reported for cores made by hydrolysis of alkoxides [10]. Losses on the order of 1 dB/km or less should be obtainable for gel-derived silica made from inexpensive commercial material.

Preparation of fibers by cast gel overcladding is shown schematically in Fig. 8. This portrays successively: dispersion, milling, casting and gelation of colloidal silica. The gel tubes formed are removed from the

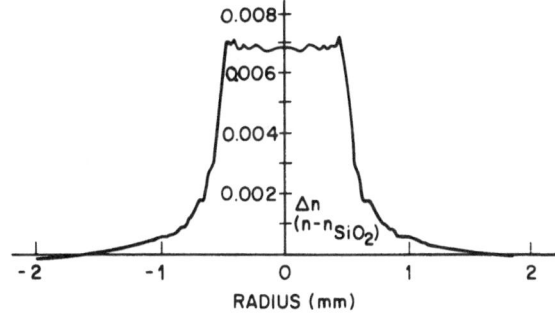

FIG. 7. Index of refraction profile for the start rod used for gel overcladding.

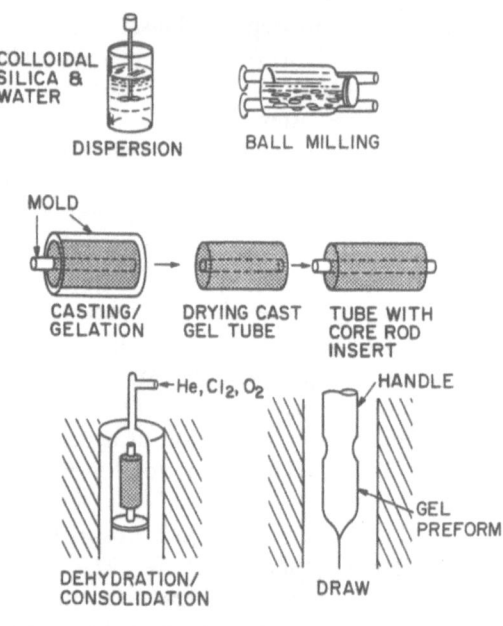

PROCESSING SEQUENCE USED FOR GEL

FIG. 8. Processing sequence used for gel overclad fibers.

mold, air-dried and placed over a core rod; the assembly is dehydrated and consolidated, drawn and coated. The core rod can be produced by vapor deposition, OVD, VAD or MCVD. It is consolidated and stretched to yield a thin rod consisting of a germanium-doped core and silica primary cladding as described previously.

Of course, a high-quality interface, free of bubbles and other distortion, between the consolidated core rod and the porous silica overcladding is required to yield low loss. By proper etching of the core rod and proper consolidation, the loss of the eventual fiber can be as low as that of the original core rod when it is drawn and measured as a multimode fiber. This is shown by the curves of Fig. 9, which plot loss vs λ^{-4}. The solid curve represents the loss of the original vapor-deposited rod and the dashed curve is that of the overclad fiber. Slight differences between the two curves result more from the better confinement of the optical power by the multimode core than from absorption by the overcladding.

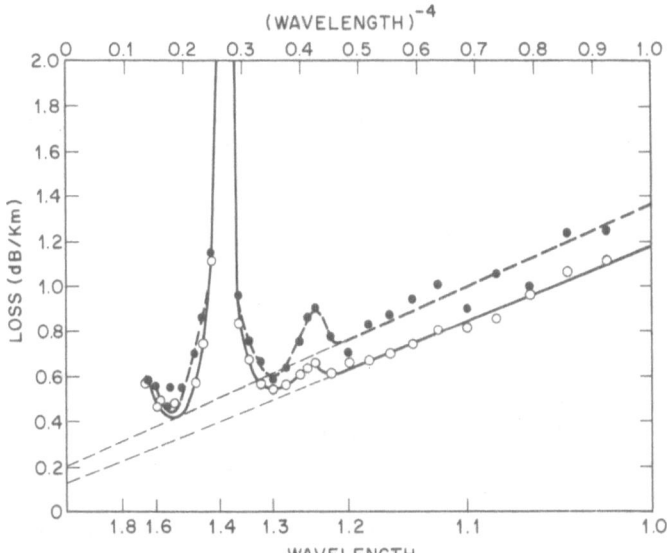

FIG. 9. Comparison of loss vs wavelength^{-4} for initial core rod drawn as a multimode fiber (solid curve) and single-mode fiber formed by overcladding same rod (dashed curve).

These experiments demonstrate that near state-of-the-art losses can be obtained by both all-gel fibers and by gel overcladding. Fibers consisting of 95% gel-derived and 5% vapor-deposited silica can be made from inexpensive fumed silica. This material costs less than 1 U.S. cent/gram and can be fabricated easily into gel bodies to provide the overcladding using simple equipment without need of extraordinary means to maintain purity. In fact, impurities originally present in the starting material and those added during processing are dramatically reduced by equilibration with a chlorine-containing atmosphere.

This chapter has reviewed a number of approaches to the preparation of optical fiber from commercial precursors: TEOS, SiCl$_4$, or fumed silica. Fiber exhibiting acceptable properties has been demonstrated and prospects are that improvements will occur with further development. We have prepared larger tubes (19 mm × 40 mm × 60 cm, weighing up to 1·1 kg) and have overclad MCVD preforms with them. The strength of fiber drawn from these meets present standards.

Still much work must be done before these demonstrations can lead to
a technology competitive with the mature vapor deposition processes
currently in use.

At this point one cannot foretell with certainty that sol gel will gain
a dominant position in waveguide or glass processing. Still its addition
to this evolving materials technology is an interesting one. Original
attempts sought to use conventional glass processing employing highly
purified solid constituents to make waveguides. This failed because of
an unavoidable and an unacceptable contamination during the processing. Next, successful product was made using intrinsically pure vapor
phase constituents with the added benefit that purification continued
during the processing. From this experience, it was learned that
purification of porous silica bodies could be accomplished by equilibration at high temperatures in chlorine. When understanding of the
role of oxygen partial pressure was added to this knowledge, processing of impure (and inexpensive) colloidal material to waveguide
quality vitreous silica becomes possible. We have demonstrated that
transition metal ions, alkalis as well as hydroxyl impurities present in
the starting material, as well as those added during mixing, grinding,
casting, etc., will be removed before consolidation. Thus there seems
to be no 'fatal' flaw to a sol–gel processing strategy which forecloses its
continued development.

REFERENCES

1. A. D. Pearson and W. G. French, Low loss glass fibers for optical
   transmission, *Bell Labs. Record,* **50,** 103–6 (1972).
2. D. B. Keck, P. C. Schultz and F. Zimar, U.S. Patent 3,737,292 (1973).
3. J. B. MacChesney, D. B. O'Conner, F. V. DiMarcello, J. B. Simpson and
   P. D. Lazay, Preparation of low loss optical fibers using simultaneous
   vapor phase deposition and fusion, *Xth Int. Congress on Glass,* Kyoto,
   Japan, 1974, pp. 6–40.
4. T. Izawa and N. Inagaki, Materials and processes of optical fiber
   fabrication *Proc. IEEE,* 1184–7 (1980).
5. J. Zarzycki, M. Prassas and J. Phalippou, Synthesis of glasses from gels:
   The problem of monolithic gels, *J. Mater. Sci.,* **17,** 3371 (1982).
6. J. O. van Lierop, A. Huizing, W. C. P. M. Meerman and C. A. M.
   Mulder, Preparation of dried monolithic SiO_2 gel bodies by an autoclave
   process. *J. of Non-Cryst. Solids,* **82,** 265–70 (1986).
7. R. Dorn, A. Baumgartner, A. Gutu-Nelle, W. R. Rehm, S. Schneider
   and H. Haupt, Glass fibers from mechanically shaped preforms, *Glastch
   Ber.,* **66,** 79–82 (1987).

8. J. B. MacChesney, D. W. Johnson, Jr., D. A. Fleming, F. W. Walz and T. Y. Kometani, Influence of dehydration/sintering conditions on the distribution of impurities in sol–gel derived silica glass, *Mater. Res. Bull.*, **22**, 1209–16 (1987).

9. S. Shibata and M. Nakahara, Low-OH-content fiber fabrication using particle-size control sol–gel method, *Tech. Digest, 11th European Conf. Opt. Comm.*, Venice, Italy, 1985, pp. 1–2.

10. T. Kitagawa, S. Shibata and M. Horiguchi, Wholly synthesized fluorine-doped optical fibers by sol–gel method, *Electronics Lett.*, **23**, 1925–6 (1987).

11. J. W. Fleming, T. J. Miller and R. E. Jaeger, GeO_2–B_2O_3–SiO_2 optical glass and lightguides, U.S. Patent 4,011,066 (1977).

12. J. B. MacChesney, D. W. Johnson, Jr., D. A. Fleming and F. W. Walz, Hybridized sol–gel process for optical fibers, *Electronics Lett.*, **23**, 1005–6 (1987).

13. J. W. Fleming, Sol-gel techniques for lightwave applications, *Tech. Conf. on Optical Fiber Comm.*, Reno, Nevada 1987, Paper MH-1.

14. S. Sudo, M. Nakahara and N. Inagaki, A novel high rate fabrication process for optical preforms, *Tech. Digest, Fourth Int. Conf. on Integ. Optics and Opt. Fiber Comm.*, Tokyo 1983, 27 A 3–4.

8

Advances in the Performance of Cement-Based Systems

F. P. GLASSER

*Department of Chemistry, University of Aberdeen,
Old Aberdeen, UK*

ABSTRACT

*Despite more than a century of research, considerable potential exists
for improving the performance of cement-based systems. The performance goals have also shifted in priority. Durability is now viewed as
having higher priority than strength, except in certain special applications. In this chapter, emphasis is placed on the chemical and
microstructural features of cements and blended cements. The engineering properties of cement materials can be measured independently of
these features, but if engineering properties are to be explained, and
their sensitivity to environmentally-conditioned change and their response to normal ageing determined, it is essential to characterize the
chemical and microstructural features.*

*Analysis of trends in cement production shows that the drift towards
cements with higher $Ca(OH)_2$ and sulphate contents is undesirable. It
leads to subsequent changes in mineralogy and unstable
microstructures.*

*It is unlikely that totally new constructional cements will be
developed; instead, cement formulations will change in an evolutionary
manner in response to these demands.*

*The use of blending agents is seen as desirable. However, the
principles governing their most efficient use are not well understood.
Since the behaviour and performance of cement systems depend on
both intrinsic and extrinsic reactions, and the latter are conditioned by
the nature of their service environment, both equilibrium and kinetic
factors control performance.*

The interaction of kinetics with equilibria is complex: in order to understand and analyse the reactions which are occurring it is necessary to have essentially mathematical models which relate system properties to microstructure, chemistry, mineralogy, etc. While it is not at present entirely clear how such models will develop, some initial steps have already been taken with encouraging results.

1. INTRODUCTION

Cements and concretes (the latter contain cement as a binder for mineral aggregates) have evolved steadily over the past century or more. Much of the improvement in their performance arises from the application of physicochemical studies. These studies were initially most successful in characterizing the solid anhydrous calcine which, when appropriately ground, perhaps with a set retarder, is commonly known as 'Portland cement'. For the past 50 years it has been possible to explain the relationships, admittedly complex, between chemical composition, thermal cycling in the kiln, phase composition of the product and its setting and strength gain.

The Second World War, 1939–45, may also be taken as a watershed in cement research. Owing to advances in techniques, including electron microscopy and, subsequently, scanning microscopy and energy dispersive analysis, it became possible for the first time to examine in detail the microstructure and constitution of the hydrated phases. Of course, other techniques have also contributed greatly to advancing the state of knowledge of hydration processes: infrared spectroscopy, porosimetry, surface area determination, calorimetry, X-ray diffraction; studies of the naturally occurring related mineral phases, and, more recently, NMR, have all made essential contributions. It has also been possible to consolidate and integrate many of these findings, with the result that our present understanding of hydration processes, while far from perfect, is at least much improved, although still incomplete in certain respects. In part, the slow pace of advance is due to our limited ability, using presently available techniques and theories, to characterize amorphous or nearly amorphous solids. In this respect, cement science shares characterization problems with other related areas of materials science, e.g. those dealing with glass science, gel technology, etc.; indeed, wherever indirect characterization techniques need to be applied in order to determine the structural state of amorphous materials.

There is a comfortable, if unwarranted, tendency for science planners to regard cements as fully developed materials which are unlikely to benefit significantly from further research time or expenditure. Little justification exists for this view, for a number of reasons, both technical and economic.

The socio-economic factors will be explored only briefly. These centre on the energy component of cement making. Steep increases in the cost of energy in the 1970s provided stimulus to short-term research into ways of reducing the energy component of cement making. This has led to the phasing out or conversion of old, inefficient kilns, often using the wet process, and to the widespread introduction of suspension preheating. The overall energy consumption for cement making has dropped substantially as a result of these changes. However, the limits imposed by the thermochemical cycle of cement making still remain; cements require a substantial energy expenditure, much of which is irrecoverable, as it is stored in the clinker minerals where it subsequently provides the activation energy for hydration. Figure 1 presents this thermodynamic cycle, showing why the energy content of clinker must remain high: the clinker phases have to remain highly metastable with respect to their hydration products in order to facilitate reaction. The diagram is schematic, because the system is open to addition or removal of certain

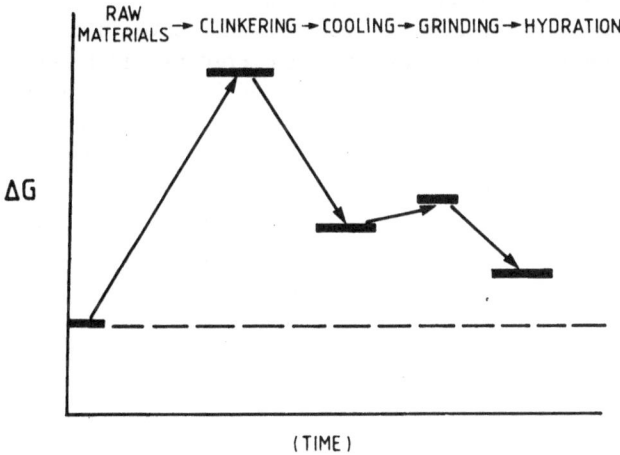

FIG. 1. Schematic diagram showing free-energy changes occurring during the cycle of cement making and hydration.

components, e.g. loss of CO_2 which occurs during clinkering, but it is nevertheless useful for inductive thinking. The high-energy state of the clinker is sufficient to ensure that kinetic barriers to hydration are overcome, and that the clinker will hydrate readily with water to form a bonding phase or phases.

Raw material substitution has also increased. Ground granulated blast furnace slag and certain glassy coal combustion products such as fly-ash now find increasing use as cement blending agents, where they partially replace Portland cement.

Energy prices have, however, fallen over the past decade, with the result that further efforts to economize have slowed. The respite is, however, only temporary. As world supplies of energy become exhausted, energy prices will again begin to move upwards more rapidly than the general cost of living. Oil and gas will be more sensitive price indicators than coal, so cement production costs will tend to rise most rapidly in countries which have no indigenous coal reserves to exploit and, moreover, have a chronic balance-of-payments deficit.

It is often and somewhat cynically argued that, as cement competes with steel, the latter (as well as other structural materials) will undergo even more rapid price rises relative to cement, on account of their higher energy contents, thus leaving cement, on balance, still competitive. However, the history of technology change shows that no material is forever immune to competition.

Moreover, cement–concrete constructions have often proved to have but limited serviceability, and, as energy costs increase, they will inevitably suffer by comparison with steel. In less-developed nations, cheap cementing materials are required in ever-increasing amounts, as the indigenous populations attempt to build an infrastructure comparable to that which advanced nations presently enjoy.

The problems associated with devising cheap, durable and environmentally acceptable materials is clearly one which is not readily solved. While it is *possible* that entirely new materials will be developed during the next few decades, this scenario seems unlikely for several reasons. One is the long time scale required to prove that a new structural material will continue to perform for long periods. This in-built conservatism of engineers and architects, marked by unwillingness to adopt new, unproven materials, is not without justification: the history of building technology is marked by failures of new 'wonder materials'. Hence the emphasis on using existing, proven

TABLE 1
Target areas for developments in cement and concrete technology

Specific area	Action for implementation
Improved fuel economy	New kiln designs to facilitate heat transfer: 'low-energy' cements, utilization of partly-processed waste materials as kiln feeds, etc.
Granulometry	Improved grinding technology, leading to better cement powder packing; better aggregate grading
Rheology	More scientific understanding of the behaviour of water-reducing additives
Improved range of mechanical properties	Novel processing techniques, e.g. MDF materials to eliminate pores, better microstructural control of hydrated product
Better composites	Understanding the role of both low-modulus (e.g. cellulose) and high-modulus (e.g. steel) fibres as reinforcements: development of fracture-mechanical principles
Application of blending agents	More characterization data for PFA, slag, etc.; improved understanding of kinetics and mechanism of hydration
Durability	Better quantification of the nature and rate of environmental response of cement systems
Information technology	Transfer of findings of research studies to engineers and their implementation in the form of codes of practice

materials whose least-desirable properties can hopefully be modified by research and evolutionary development, but without sacrifice of their desirable properties.

There is a significant list of areas in which advances in cement manufacture and utilization could potentially be made. Table 1 gives some of these. It may be noted that some of these potential areas of improvement, such as improved fuel economy, rest largely on process engineering; others, such as the development of better composites, rely on improvements to the theory and better application of fracture mechanics to brittle materials, while still others, such as improved durability, rest on interdisciplinary advances. The extensive use of natural and synthetic low-modulus fibres, e.g. cellulose or PVA, as replacements for asbestos creates problems in understanding the mechanical properties of the resulting composites and of their ageing behaviour.

The selection of examples chosen for further exposition in this presentation is, inevitably, one of personal choice; not all relevant themes will be developed equally. Indeed, it seems profitable to approach these themes broadly, by examining the *performance* of cement-based materials. One such approach, integrating several of the themes in Table 1, is to examine the *durability* of cemented materials.

2. DURABILITY OF CEMENTS AND CEMENT COMPOSITES

2.1. Scope

It must be stated at the outset of any consideration of durability that the inadequate performance of cement-based systems can arise from a wide variety of causes. Several recent seminars [1–3] have highlighted a number of causes and given descriptions of the principal mechanisms of deterioration. Inadequate design, poor construction practice and lack of attention to materials selection play an important part in determining the durability of cement matrices and the serviceability of structures. However, cement matrices have certain intrinsic physical and chemical features which cause them to differ from almost all commonly used constructional materials. This makes comparison with other solid materials difficult; for example, with metals; exact parallels in behaviour frequently do not exist, nor is it possible to apply quality control procedures developed for metals to cement-based materials.

Durability assumes increasing importance in the developed nations, where experience shows that, increasingly, a long life is expected from constructions. The high cost of construction, and the even higher cost of reconstruction, on account of its higher labour content, means that the real cost–benefit to owners, occupiers and the community generally is enhanced if the structural components of the built environment have a long life.

While these considerations are increasingly accepted as very relevant, they have not always worked through so as to affect design and construction, although they may be expected increasingly to do so in the near future.

TABLE 2
Durability of building materials and components applied to cements

Classification	Subdivision	Example
Equilibrium changes related to thermodynamic potentials	Intrinsic	Reaction of cement alkali with certain silica; e.g. opal.
	Extrinsic	Uptake of CO_2 leading to carbonation. Also, special hazards, e.g. sugar solutions, molten metals
Equilibrium changes related to kinetic potentials	Intrinsic	Specific design constraints imposed by formulation, e.g. use of blending agents
	Extrinsic	Response to environmental factors conditioned by concrete properties, e.g. permeability

Like most constructional materials, cements are used in a wide range of environments, where they are also exposed, at least potentially, to a wide range of degradation processes.

In general, the durability of concrete materials in service consists of two aspects, shown in Table 2. The first set of reactions arise as a consequence of catastrophic events, for example, fire, earthquake, etc. Such risks can only be assessed on the basis of a probabilistic model, and a response to the assessment made in terms of design and quality control. The area of catastrophic events will not be addressed, although it is clearly one of importance. However, the integrity of structures, and their ability to withstand unusual stresses, may decline as a function of time owing to degradative processes. These constitute the second set of mechanisms, which will be addressed here. The degradation of materials occurs, broadly, as a consequence of the laws of thermodynamics, and at a rate governed by the kinetics of thermodynamically favoured specific processes. The applied stress (chemical, physical) usually occurs at a more constant rate than catastrophic events, although periodic or episodic events are not precluded, e.g. freeze–thaw. Moreover, the nature of degradative processes is such that different mechanisms may interact with each other.

The separation of degradative processes into two classes enables us to analyse the mechanisms of deterioration. The *equilibrium* aspects relate to the in-built tendency of the system to reach a physicochemical equilibrium between its various components. Thus, siliceous materials which may be added to cement as aggregates have an in-built thermodynamic potential for reaction with the alkaline components of cement. This is termed an intrinsic thermodynamic potential, which, in this instance, would be conditioned by the original constitution of the system. Extrinsic potentials, on the other hand, arise from environmentally conditioned reactions involving transfer of material into or out of (or both) the cement system.

A favourable thermodynamic potential for either intrinsic or extrinsic reaction is not of itself sufficient to ensure that reaction will occur over the time scale of observational experience, typically $\sim 10^2$ years. For example, all siliceous aggregates, including well-crystallized quartz, are probably alkali-susceptible and provide, at least in theory, intrinsic potential. But the rate of reaction, broadly defined by the term 'kinetics', is often sufficiently slow as to be imperceptible on this time scale with the result that only certain reactive silicas, such as tridymite, cristobalite or opal prove to be troublesome.

This proposed classification, having as its basis thermodynamics or kinetics, may not be convenient or comfortable for conventional engineering use, but it does provide a link between engineering observation and laboratory studies, whereby degradative potentials can be characterized and measured. Moreover, unless measurements can be made in the context of a physicochemical framework, as is imposed by thermodynamic and kinetic approaches, durability cannot be quantified. But, since the cement component is only one of the materials which together form the complex composite chemical and mechanical structure known as 'concrete', it is also necessary to consider additional factors.

2.2. Concretes as Composite Materials

Concretes should also be treated as composite materials. Table 3 illustrates some components of these composites. The cement matrix consists largely of a high-surface-area, non-crystalline inorganic gel, containing principally CaO, SiO_2 and H_2O. Since this gel, as well as much of the alkali content of the cement, is appreciably soluble, mechanism exists for material transport whereby the substantial chemical potential between cement and other components of the composite system can be reduced. Reaction is achieved most rapidly in

TABLE 3
Concretes as composite materials

Composite component	Reaction with cement
Steel	Partly passivated by reaction with cement at high pH; slow corrosion due to H_2 evolution
Glass	React with alkaline cements; glass is hydrolysed and fibre integrity is lost
Organic fibre	Properties observed to deteriorate with time, but mechanism(s) not well understood.
Mineral aggregate	
Siliceous components	React with alkaline pore fluid of cements, creating expansive moment
Dolomite component	React with alkaline components of cement: mechanism not well known
Layer–lattice component	Shrink–swell, arising from ion exchange, wet-dry cycling, etc.
Water	Freeze–thaw: physical damage

the presence of water; not only is this water rather loosely held in the gel phase, but excess water, beyond that required for hydration of the cement phases, is normally present. The soluble cement species cause this excess water to become strongly alkaline. The alkaline nature of these cement pore fluids in turn conditions many of the interactions which are observed to occur with other components of the composite. The internal chemistry, as measured by the pH function is never less than 12·4, but may rise to nearly 14. The internal environment conditioned by the fluid phase has both desirable and undesirable consequences. An example of the latter has already been adduced: it facilitates reaction with certain alkali-susceptible aggregates with glass-fibre reinforcement, etc. However it may also have beneficial aspects:—it creates a chemical environment leading to passivation of embedded steel reinforcements. The matrix may not persist unchanged in its service environment, but may instead undergo environmentally conditioned reactions such as those shown in Table 2, which lead in the long term to changes in its properties. Considerable care is required, therefore, to determine the properties of composite systems at long ages: it is almost certainly incorrect to conclude their properties on the basis of short-term measurements, typically obtained at less than 1–2 years. It is in this area that we concentrate initially, as

TABLE 4
Characterization parameters relevant to durability

Factor	Typical characterization data
Chemical	Bulk solid composition, pore fluid compositions
Mineralogical	Phase content and amount of constituent phases
Physical	Microstructure; relation of crystalline phases to each other; amount and connectivity of porosity; permeability; microcracking

the changes to matrix properties are often rate- and process-determinant.

Characterization of the matrix can take different forms, some of which are shown in Table 4. Much has been written about the identity and characterization of the constituent phases, but it is intended to concentrate here on the microstructural relations, about which less is known.

2.3. Microstructural Relationships

With the advent of electron microscopy, in conjunction with optical microscopy, it has been possible to define microstructure across the relevant scales. These extend from several centimetres in the case of aggregates, right down to the nanometre level, as in the case of paste structures. We will, however, be concerned with the smaller-scale features occurring in pastes. This level of microstructural exploration has occurred at both qualitative and quantitative levels. Quantitative descriptions have had to await the introduction of image analysis, which is relatively recent, so the flow of information from this source has yet to make as full an impact as has more qualitative information. We can, however, look forward to obtaining much more information in the future: moreover, quantitive microscopical data should complement and reinforce data obtained from other sources, e.g. intrusion porosimetry. The microstructures of cement systems differ sharply from those of other systems—for example, those of polymers, metals and glass—so that extension of the microstructural quantification may well involve more than a straightforward application of well-established techniques to a new material. Nevertheless, sufficient is now known about the microstructure of cements to be certain that, in common with many other materials, microstructure plays a role in determining many aspects of their durability.

Thus, for the first time, we are in a position where it should be possible to quantify cement microstructure. The potential benefits which may thus accrue can perhaps best be appreciated by comparison with other areas of material science, where quantification of the microstructure has proved to be the key to its control. This leads to the expectation that great advances will occur over the next few decades first in the characterization and then in the control and optimization of the microstructure of cement-based products. This will in turn lead to enhancement of their properties and performance.

2.4. Origin and Description of the Microstructure

During hydration, cement clinker grains are converted to hydrated solids. While some disagreement exists concerning the relative importance of solution-and-reprecipitation reactions versus direct, in-situ hydration, these arguments are now largely historic. Cement chemists would accept that both mechanisms are important: during the early stages of hydration, while much fluid water still remains, solution–precipitation reactions are important. At longer ages, beyond a few days, the matrix becomes increasingly solid-like, so that liquid water accumulations are increasingly confined to pores on a micrometre scale and the importance of in-situ hydration reactions increases. Thus, the term 'outer product' refers largely to material formed by through-solution hydrates; these comprise the bulk of the early hydration product, together with material subsequently formed in the vicinity of fluid-filled pores. Inner hydrate forms as a result of in-situ conversion of clinker phases to hydrates as H_2O and OH^- diffuse through hydrate product layers towards the remaining kernels of unreacted clinker phases. Since set occurs relatively early in the history of a cement, it would appear that from the mass balance standpoint, most of the hydration process occurs by in-situ hydration. Although the wide variety of slightly divergent interpretations offered in the literature may not appear to converge on a single interpretation, any remaining areas of disagreement should not be allowed to obscure broad agreement: this view of cement microstructure is believed to be supported by literally hundreds of papers reporting optical and electron-optical examinations. Thus, hydration for short times, up to ~1 year at low temperatures (~20°C), gives rise to consistently observed microstructural features, the origin of which relate closely to the packing of cement particles and to the water : cement ratio. During hydration, partial filling of the larger pores occurs by growth of hydrates into space formerly occupied mainly by aqueous phase, so that hydrates formed by relatively unrestrained conditions exhibit morphologies characteristic of rapid growth from solution. On the other hand, the inner hydrates, growth of which is mechanically restrained, are relatively featureless but are also free of larger pores.

It therefore follows that the morphological structures occurring in cements range over a wide range of sizes and scales. Figure 2 illustrates some of these microstructural features and their scale. Of the raw materials used in preparing cement paste, the median particle size tends to decrease in the order clinker > fly ash > silica fume, with the relative position of slag particles dependent on its grinding history but typically, it is slightly finer than cement. The clinker grains are themselves polycrystalline and comprise both relatively coarse crystallites, ranging up to several micrometres, of both C_3S and C_2S, together with smaller crystallites, often with dendritic habits, of C_3A, ferrite, etc. Water molecules, on the other hand, are many orders of magnitude smaller than even the finest solid particulate matter. Owing to the nature of the particle packing, conditioned by the cement/blending-agent/water mix, the spaces available for accumulation of the 'through solution' hydration products, such as large portlandite crystals, outer hydrate, etc., are essentially fixed because these newly formed solids preferentially occupy the sites of what were originally water-filled voids. However, the total volume of solid

CHARACTERISTIC SIZE RANGE	< 1 nm – 10 nm	< 10 nm – 100 nm	< 100 nm – 1 μm	1 μm > 10 μm
REACTANTS	WATER MOLECULE		SILICA FUME	FLY ASH, CLINKER, SLAG
SOLID REACTION PRODUCTS	ORDERED C-S-H REGIONS		C-S-H 'STRUCTURES' AF_m	$Ca(OH)_2$ (Portlandite) AF_t
VOIDS, ETC	GEL POROSITY, CHANNELS IN AF_t, ETC.	MESOPOROSITY		MACROPOROSITY BUBBLES, MICROCRACKS ETC

Fig. 2. Schematic diagram illustrating the scale of microstructures occurring in cement pastes.

hydrate developed is generally insufficient to fill these voids completely, so some remain as open porosity. One obvious source of microstructural enhancement, therefore, has as its aim the more complete filling of interparticle voids; hence the importance of ensuring maximum densification of the powder packing, consistent with being able to retain sufficient water in interparticle spaces for hydration. This least requirement is, however, not very restrictive in normal practice, since the critical water : cement ratio required for complete hydration is only about 0·22–0·24. A wide range of organic plasticizers can be used to maintain an appropriate rheology of fresh cement, but at lower water contents than would otherwise be required.

The hydrate phases also characteristically exhibit 'structure' over a very wide range of sizes. The contrast between the ordered regions of C–S–H and its particle size is noteworthy; although C–S–H often appears to have well developed morphology (fibrils, foils, etc. are reported) these larger-scale C–S–H features arise from an essentially amorphous material: C–S–H exhibits ordering characteristic of crystalline substances only on a nanometre scale. The spaces between the irregular stacking of the more ordered nanometer-sized units contain much of the so-called gel porosity. AF_m, AF_t and $Ca(OH)_2$ are, however, rather better crystallized. These characteristic differences between phases need to be taken into account when engineering the microstructure: not all phases are equally suitable to act as pore fillers, and the C–S–H, although amorphous, still contributes more to the strength than do crystalline phases such as AF_t.

One microstructural feature of cements which is very striking, and needs further exploration, is the comparatively unstable nature of the normal microstructure which has been described here. Typical characterization data in the literature relate to cement pastes made to 'normal' water : cement ratios (approximately 0·3–0·5), aged up to ~1–2 years at moderate temperatures, typically 18–25°C. However, the microstructure thus obtained is, in fact, not very stable and is particularly sensitive to temperature. Some of the effects of thermal exposure can be appreciated by comparing the micrographs shown in Figs 3 and 4, taken from Ref. 4 of identical cements after different thermal treatments. Both cements, to BS12 (very similar to ASTM Type 1), were cast from one batch in ~200 g 'Perspex' cylinders at a w/c ratio of 0·35. After 24 h cure in sealed molds at ~20°C, the cylinders were demolded and subjected to different cure regimes; 25°C, 40°C and 55°C for 180 days at 100% humidity in a CO_2-free

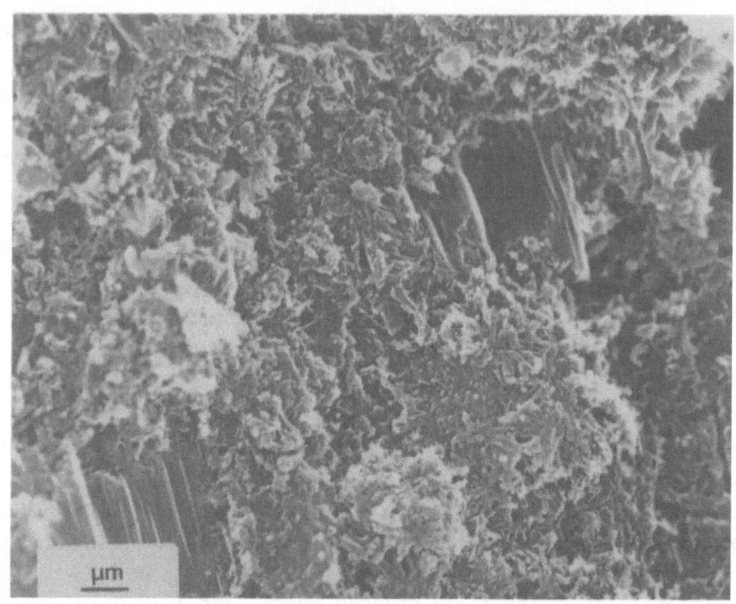

FIG. 3. Effect of ageing on cement microstructures: 180 days at 100% humidity, 25°C.

FIG. 4. Effect of ageing on cement microstructures: 180 days at 100% humidity, 55°C. [4]

atmosphere. After cure, they present a very similar appearance, except for the scale of their microstructures, which coarsens progressively as the cure temperature is raised, more at 55°C than at 25°C. The two micrographs selected for illustration can be compared directly, as they were taken at identical magnifications. The coarsening of the pores at 55°C was also demonstrated by mercury intrusion porosimetry. The relevant porosity curves are shown in Figs 5 and 6, from which it can be seen that although the total intruded volume is about the same in both cases, the 55°C cure shifts the pore size distribution significantly to a coarser regime. Since intruded pores are also accessible to diffusing species, this implies that the high-temperature cure results in a higher permeability. Supplementary thermogravimetric analysis (TGA) shows that both contain about the same total quantity of $Ca(OH)_2$, but X-ray diffraction measurements disclose significant changes in mineralogy: AF_t, present in the 25°C cure, was absent at 55°C. The 40°C cure gave microstructural coarsening intermediate between that at 25°C and 55°C. At 40°C, AF_t, while present, was reduced in amount relative to 25°C.

What are the implications for durability? There are several. Firstly, the microstructure characteristic of 'normal' cure regimes is unstable. While cement microstructures may coarsen only slowly at low temperatures, up to 25°C, they can be expected to coarsen much more

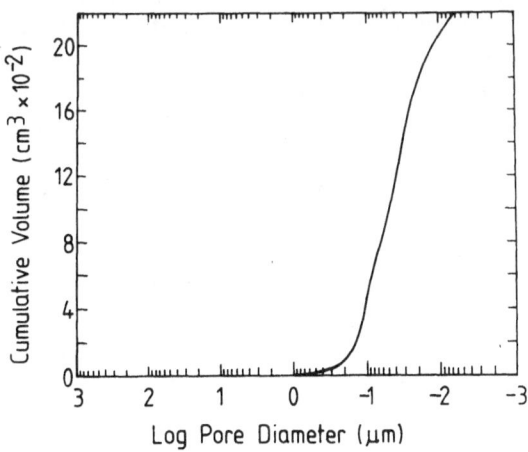

FIG. 5. Mercury intrusion porosimetry data on the cement whose microstructure is shown in Fig. 3.

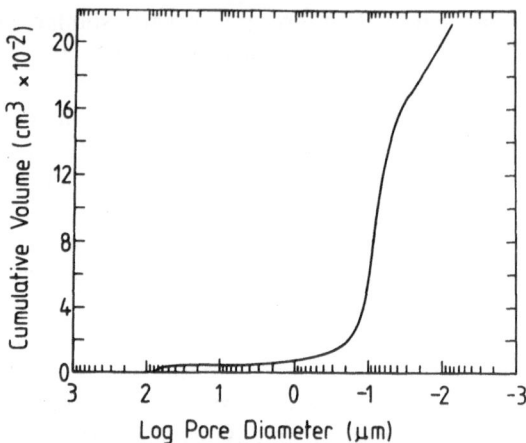

FIG. 6. Mercury intrusion porosimetry data on the cement whose microstructure is shown in Fig. 4.

rapidly as the temperature rises to 40°C and 55°C. Coarsening of the microstructure increases mean pore sizes and, presumably, enhances permeability; it is also associated with significant changes in mineralogy. Secondly, the changes arising from thermal exposure probably have both a reversible and an irreversible component. The coarsening of the microstructure is believed to be irreversible, but the conversion of AF_m to AF_t is believed to be reversible at least in part, as will be explained.

These extensive mineralogical and microstructural changes imply that much diffusion is occurring. The 40°C time–temperature exposure, together with moist conditions, can realistically be achieved by concrete structures. So the typical microstructure of pastes which represent examples of current good practice must be regarded as rather unstable and, even in the absence of further environmentally conditioned reactions, liable to undergo change—slowly at 25°C, but perceptible even in 6 months at 40°C. The reversible component of the change is also potentially of great importance. Generally, high-temperature cure favours conversion of AF_t to AF_m, while at lower temperatures the direction of reaction is reversed. It is likely that no sharp thermal boundary exists between the ranges of stability of the two phases (although this cannot be ruled out) but instead that a short range exists over which, with rising temperatures, AF_t gives way to

AF_m. The conversion of AF_t to AF_m decreases the specific volume of the solid phases, a process akin to that of 'conversion' in high-alumina cements where, it will be recalled, formation of Ca_3AH_6 reduces the specific volume of the solid phases, thereby opening up microcracks. A parallel process exists, it is suggested, in Portland cements. Experience shows that in the reverse reaction, newly-formed AF_t does not passively refill pores, as might be expected of a soft gel, but instead nucleates selectively and continues to grow as lath-shaped crystals whose growth mechanism in the solid hydrates creates expansive forces within the paste.

The effects of thermal cycling are further explored, admittedly in a somewhat speculative manner, in Fig. 7. Two pathways are shown, one assuming self-heating during set, with only slow release of the heat of hydration. The other path simulates isothermal cure, followed by artificially-induced exposure at elevated temperatures. Both follow a rather similar course. All early-formed product is reconstituted and new, coarse product is formed during the high-temperature excursion.

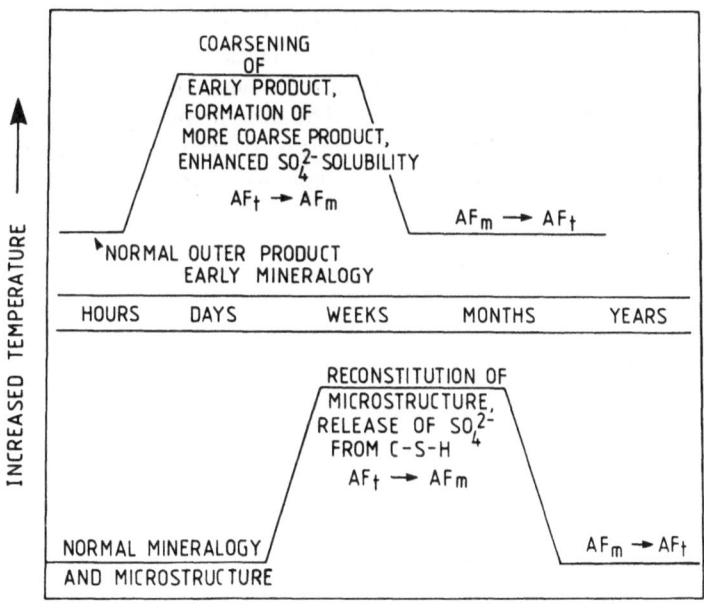

Fig. 7. Effect of thermal cycling on the microstructure and mineralogy of hydrating cement.

C–S–H which, when formed at low temperatures, sorbs a significant fraction of the sulphate ions, discharges sulphate during high-temperature exposure. AF_t phase also becomes unstable despite the higher SO_4^{2-} pore fluid concentration. However, on cooling, AF_t again becomes stable, especially as more sulphate becomes available as a consequence of its desorption from C–S–H. This sulphate is gradually removed from the pore fluid while AF_m is reconverted to AF_t, crystallization of which leads to expansive stresses and an overall decrease in strength. The extent to which late-formed ettringite can act as pore blockers is not known: there could be a potential benefit arising from its formation, although past experience has generally shown that delayed ettringite formation is expansive, but that it is difficult to control or regulate the amount of expansion. Hence, ettringite is not likely to be suitable as a pore-blocking phase. The availability of sulphate for ettringite formation is further increased by the apparent irreversibility of sulphate sorption by C–S–H; in the course of sulphate release, C–S–H undergoes structural changes which inhibit its subsequent ability to resorb sulphate. These are currently being explored at Aberdeen using ^{29}Si NMR to determine changes in silicate polymerization.

In summary, therefore, the exothermic heat of hydration tends to generate large temperature excursions. These are indesirable. Not only can thermal cracking develop but high temperatures cause irreversible microstructural changes. The accompanying mineralogical changes are partly reversible, but the reversible element, associated with delayed ettringite formation, is likely to contribute to volumetric instability.

2.5. Relation of Chemical Composition to Mineralogy and Microstructure

Two trends in the development of modern cements are relevant to this discussion. The first is the increase which has occurred in the ratio of C_3S to C_2S in modern cements. Increasing this ratio in general increases the rate of strength gain, but, in the course of hydration, also increases the amount of $Ca(OH)_2$ which is eventually formed. Thus, many modern cements when fully hydrated contain on the order of ~25% $Ca(OH)_2$. The presence of so much $Ca(OH)_2$ facilitates coarsening of the microstructure, owing to its slight solubility and differences in surface free energy between large and small crystals. This, together with higher alkali levels, conditions the slow development of larger crystals of $Ca(OH)_2$. These crystals are mechanically weak; the layer-lattice structure of $Ca(OH)_2$ has perfect cleavage, and on that account, larger crystals of $Ca(OH)_2$ facilitate crack propagation. The second tendency is to add more sulphate to modern cements. Historically, the first Portland cements contained little if any sulphate: owing to their low C_3S and C_3A contents and their coarse grind, set retarders were largely unnecessary. However formulation of modern cements requires more retarder, and the increased sulphate contents of modern cements leads to formation of more sulphate-containing phases amongst the hydration products. As the sulphate mineralogy is somewhat unstable, more microstructural adjustments are necessary to accommodate the resulting changes: the temperature dependence of the AF_t–AF_m transition and its consequences have been discussed. For these reasons, the microstructure of modern cements is considered to be less satisfactory with respect to its potential for durability than those of early Portland cements.

Thus good factual evidence exists showing cement microstructures to be rather unstable. This is probably also true of the mineralogy. The kinetics of change are markedly affected by brief high-temperature excursions, although these kinetics are not as yet well enough determined to correlate with strength, durability, etc., although indirect evidence suggests that strong correlations will exist. To summarize, if the durability of cement matrices is to be optimized, two goals seem to be important: first, to create a favourable initial microstructure, and second to create a stable microstructure, defined as one which is resistant to change within the normal parameters governing the conditions of its service.

One potential way of creating a more stable microstructure is by adding blending agents, and this also emerges as an important area of investigation. Blending agents such as fly ash which have a relatively high silica content react with $Ca(OH)_2$, thereby shifting the bulk composition of the hydrated system closer to those of 19th-century cements. Of course, chemical reactions occur with the other components; and these may also have beneficial effects.

2.6. Pore Structure and Its Relationship to Strength and Durability

The pore structure can also be determined on dried material by mercury intrusion porosimetry. Drying, as well as the act of intrusion, may result in some damage to the paste structure, but in general this affects mainly the sub-micrometre structural features, so that mercury intrusion porosimetry provides a reasonable guide to the pore distribution and connectivity, especially of the larger macro- and mesopores. The porosimetry confirms what would also be inferred from diffusional measurements on pastes: that as their water : cement ratios increase, the amount of large porosity increases rapidly. Furthermore, many of the larger pores are seen to be interconnected.

The relation between compressive strength and water : solid ratios was established by Feret as early as 1897. Feret related the compressive strength, σ_t, to the volumes of cement, V_c, water V_w and of air V_a by the following equation:

$$\sigma_c = A\left[\frac{V_c}{V_c + V_w + V_a}\right]^2$$

where A is a constant. The equation is empirical, and clearly will not hold over all values of V_w; in any event, practical difficulties in ensuring homogeneous mixing, and the essential role of water in the hydration process, ensure that the ratio of water : cement has to be greater than 0·24 (approximately), for the relationship to be applicable. It has now been possible to essentially restate the volume relationships in terms of porosity, so that the strength, σ_c, at some porosity, p, is related to the strength of an ideal dense material, σ_0, by the relation

$$\sigma_c = \sigma_0 \exp(-bp)$$

Extrapolation of data on a p versus σ_c plot suggests that cements have an intrinsic compressive strength of $\sim$130 MPa and flexural strength of $\sim$20 MPa. These and other proposed relationships relating strength to microstructure are discussed in Ref. 5. It emerges from these descriptions that cements are intrinsically weak materials. Indeed, this could be anticipated from the nature of the hydrogen-bonded network of the gel phase; hydrogen bonds are intrinsically weak, and the energy required to extend the bonds is low, so cements cannot be expected to achieve the strength potential of metals or even of ceramics, whose ultimate strengths are based on metallic or partial ionic bonds, respectively. Thus, theoretical expectations of cement bond strengths are approximately in line with observed strengths. But although cements are intrinsically weak materials, their actual performance is nevertheless usually well short of that which is potentially attainable, owing principally to their high porosity. In practice, even the expected large reduction in strength arising from the presence of roughly spherical pores may be exceeded if the geometric nature of the pores is essentially crack-like.

It therefore follows that reduction in porosity is an important goal. Several methods have been used to achieve this; for example, hot pressing has been used to develop perhaps the first really high-strength cements [6]. However, hot pressing is slow and limited in flexibility. The use of superplasticizers, in conjunction with low water : solid ratios and mechanical working, has led to so-called MDF cements, with strengths ranging up to 150 MPa [7]. The forming procedures which are required to make MDF cements although simpler than hot pressing, are not practicable for use with constructional concretes, although some building components might lend themselves to pre-fabrication by MDF routes. The existence of these routes to high-strength materials does, however, demonstrate the importance of porosity in controlling the mechanical properties of the matrix.

Blended cements, containing slag or fly ash, can be proportioned so that, after adequate cure, they have porosities slightly lower than corresponding formulations with Portland only. The reduction in porosity is not normally sufficient to give significant enhancement in strength, but the reduction in permeability achieved by the presence of a blending agent is quite marked. Apparently, the hydration products of the composite formulation tend to develop in a way which blocks pore entrances. The membranes thus created are rather fragile, because mercury intrusion porosimetry tends to break them down. But they are sufficient to reduce significantly diffusion rates of aggressive ions (e.g. Cl^-) through the paste matrix. One reason why blending agents such as slag and fly-ash do not result in appreciable strength gains is that their particle-size distributions are too close to those of cement particles. The resulting particle packing is un-favourable with respect to creating a relatively dense network the interstices of which can be completely filled with hydration products. However, silica fume has a particle size distribution which is typically much finer than those of cements or fly-ash or ground granulated blast furnace slags, and it is thus well suited to forming a dense mass when mixed with Portland cement. Typically, silica fume has a particle diameter of about $0.1 \, \mu m$, some two orders of magnitude less than an average cement clinker grain. When the cement–silica–fume–water system is flocculated, relatively dense packing occurs spontaneously, typically at water : (cement + microsilica) ratios between 0.12 and 0.22. There low water content mixes require a plasticizer to be added. Depending on water content and blend ratio, the consistency of the freshly-made systems range from free-flowing to plastic. Thus, elabor-

ate compaction and forming procedures may not be required to
produce high-strength materials, and compressive strengths up to
270 MPa have been achieved [8].

Increases in the compressive strength of cement-based systems are
not, however, viewed as a high priority goal for cement systems.
High-strength materials might extend the range of applications of
cemented systems, but, for constructional uses, cements already have
sufficient compressive strengths, although improved flexural strengths
would be desirable. Much more important, however, are the potential
improvements in durability resulting from decreased porosity. In many
respects this is also easier to achieve than the stringent standards
required for high-strength materials: whereas high strength materials
require to have near-zero porosities, significant improvement in
durability, particularly in resistance to penetration by chloride, etc., is
obtained when porosities fall below ~20%. Porosities in this range,
just below 20%, are much easier to achieve than the near-zero
porosities required for high-strength materials.

Microstructural studies of cement systems disclose a number of
significant clues with respect to the future development of cement
materials by control of porosity. The first is that 'ordinary' blends,
made using commercially available cements and blending agents, have
rather high porosities, on the order of 25–30%. At these high
porosities it is almost inevitable that much of the porosity is interconnected, and interconnected porosity is associated with poor durability.
Chemical fluxes occur readily through the open pore network; CO_2,
oxygen, and chloride, etc. ingress in appropriate environments, while
leaching, mainly of alkalis, also occurs readily. But as the porosity
decreases below roughly 20%, the pore network shifts abruptly from a
regime having a high degree of interconnectivity to one with a much
lower degree of connectivity, so that diffusional fluxes are much
inhibited. The critical value of around 20% is not fixed, but appears to
depend upon the precise granulometry of the raw materials used and,
of course, the water : solid ratio. The use of blending agents, together
with thorough curing, is often sufficient to achieve the transition
between largely open and largely closed pore regimes. Unfortunately,
the 'rules' for proportioning blends to achieve low porosity, much of
which is closed, are only known empirically. Until the granulometry of
the raw materials is better controlled, and until mineralogical and
microstructural changes occurring during hydration are better understood, proportioning formulae for impervious concretes must remain

empirical. There is little doubt however, that many 'lean' concretes containing perhaps 275–325 kg cement per cubic metre are well below the desirable limit for best durability, although they may have adequate strength. Total cement content is thus another important parameter which needs to be assessed in the context of durability [9]. Low cement contents may give adequate strength, but economy in cement content may be achieved at the expense of a highly permeable matrix.

The use of silica fume is helpful in achieving dense, impervious and strong pastes, but silica fume is produced world-wide on a limited basis, whereas slags and fly-ash are more abundant and, moreover, widely available. Thus, considerable scope exists for improvement in the granulometry of cement blends, using 'common' blending agents, in conjunction with adequate total content of cementitious material, with a view to enhancing durability. Because many of the blending agents are themselves industrial by-products, more characterization of fly ash and slag are also needed, although recent state-of-the-art symposia suggest that many of the basic data are available [10–12]. In this respect, obtaining more characterization data is arguably less important than asking what type of characterization data are needed, and why. However, many of the processing techniques used for ceramics can also be applied to cements. Undoubtedly, cement systems are still capable of much development : improvements in mechanical properties and durability have been singled out as the most promising fields.

REFERENCES

1. P. J. Sereda and G. G. Litvan (eds), *Durability of building materials and components* (Proc. 1st Intl. Conf. Ottawa, Aug 1978). ASTM Special Publication 691, 1980.
2. *Proceedings of the 4th International Conference on the Durability of Building Materials and Components, Singapore.* Pergamon Press, Oxford, 1987.
3. A. P. Crane, *Corrosion of Reinforcement in Concrete Construction.* Ellis Horwood Ltd., Chichester, 1985.
4. K. Luke and F. P. Glasser, Effect of temperature on the hydration chemistry and durability of cement concrete. In *Proceedings of the 4th International Conference on the Durability of Building Materials and Components.* Pergamon Press, Oxford, 1987, pp. 188–95.
5. P. Hirsch *et al.*, *Technology in the 1990s: Developments in Hydraulic Cements.* Royal Society, London, 1983.

6. D. M. Roy and R. R. Gouda, Porosity–strength relation in cementitious materials with very high strengths, *J. Am. Ceram. Soc*; **56,** 549–550 1973.
7. J. D. Birchall, A. J. Howard and K. Kendall, Flexural strength and porosity of cements. *Nature, London* **289,** 388–390 (1981).
8. L. Horth, *Microsilica in Concrete,* Nordic Concrete Research Publication No. 1. The Nordic Concrete Federation, Oslo, 1982, pp. 901–8.
9. F. P. Glasser, Modelling the behaviour of cements, *Mag. Concr. Res.,* **39**(139), 58–9 (1987).
10. G. J. McCarthy, F. P. Glasser, D. M. Roy and S. Diamond (eds), *Fly Ash and Coal Conversion By-products: Characterization, Utilization and Disposal,* III. Materials Research Society Proceedings, Vol. 86, Pittsburgh, PA, 1987.
11. G. J. McCarthy, F. P. Glasser, D. M. Roy and R. T. Hemmings (eds), *Fly Ash and Coal Conversion By-products: Characterization, Utilization and Disposal,* IV. Materials Research Society Proceedings, Vol. 113, Pittsburgh, PA, 1988.
12. R. T. Hemmings, E. E. Berry, G. J. McCarthy and F. P. Glasser (eds), *Fly Ash and Coal Conversion By-products: Characterization, Utilization and Disposal,* V. Materials Research Society Proceedings, Vol. 136, Pittsburgh, PA, 1989.

9

Hydrothermal Synthesis in Acid Solutions—A Review

THE LATE A. RABENAU

Max-Planck-Institut für Festkörperforschung, Stuttgart, FRG

ABSTRACT

Hydrothermal synthesis in acid and neutral solutions using glass or quartz glass as reaction vessel plays an important role in materials synthesis. After an introduction into the matter, the physicochemical properties of the solvent water relevant to the method are discussed. Beginning with the first experiments from more than 150 years ago, the present state of the art is presented, including a discussion of hydrofluoric acid solutions employing Teflon as reaction vessel. Visual autoclaving within a broad range of temperatures and pressures allows the observation of hydrothermal reactions. The use of non-aqueous solvents is in its early stages and shows good prospects for the future. In the appendix a simple arrangement as well as an example of a transport mechanism are discussed in detail.

1. INTRODUCTION

Hydrothermal conditions (i.e. aqueous media above 100°C and 1 bar) are found in nature and numerous minerals are formed under these circumstances. Geologists and mineralogists were the first to simulate in the laboratory the conditions necessary for mineral formation and even today hydrothermal investigations play an important role in the field of geological sciences. These researchers enclosed the materials in a reaction vessel (an autoclave) which was heated to the required temperature, the pressure being dependent on the extent to which the vessel was filled.

After World War II the method was introduced into modern solid-state and materials research. The most striking example of this development can be found in the industrial preparation of quartz oscillator crystals.

The materials chosen for the vessel play an important role in hydrothermal reactions. Thus, the corrosive properties of the solution under the desired conditions as well as the danger of undesirable contamination of the reaction product must be considered.

With the exception of hydrofluoric acid, borosilicate and quartz glasses are quite stable towards neutral and acidic solutions as well as corrosive media. The redox potential of the solutions is hardly affected and the danger of contamination by metallic impurities is slight. This last point plays an especially important role if the reaction product is to be subjected to physical measurements. There are, however, pressure and temperature limitations to the use of glasses for hydrothermal syntheses in neutral or acidic media. The useful range for borosilicate glasses at higher temperatures is limited to 250–300°C. Quartz glass—despite its high softening point of around 1200°C—is only usable up to about 500°C, since at higher temperatures the walls are attacked. The maximum pressure depends on the diameter of the ampoules, the strength of the walls, the mechanical properties of the glass, and the temperature.

Glass and quartz glass have a great advantage in that, within a limited pressure and temperature range, the experiment can be directly observed. If the pressure–temperature range for visual hydrothermal synthesis is exceeded, the ampoule must be placed in an autoclave along with a pressure compensation in order to prevent explosion.

This review is limited to systems for which glass, quartz glass or Teflon are the reaction vessels, the last being necessary for hydrofluoric acid media. This limits the temperature range to about 500°C. In principle it is possible to work in acidic and corrosive media at higher temperatures; however, this requires sophisticated (metallic) apparatus, which is only available for specialized laboratories. Articles touching the subject have been published [1–4] and detailed information about relevant references are presented in the literature [2–4]. A simple method especially suited for preparative purposes and which can be used for non-specialized laboratories is given in Section 1 of the

Appendix. The role of complexes for the transport of gold from cold to hot is discussed in Section 2 of the Appendix.

2. WATER AS SOLVENT

The special aspects of the hydrothermal method stem from the physicochemical properties of the solvent water at the applied temperature and pressure, in the case to be considered here up to about 500°C. Figure 1 shows the $p-T$ diagram of pure water, with the filling factor (degree of fill) of the autoclave as a parameter. Since the filling factor is defined as the ratio of the volume of liquid to the volume of the reaction vessel (at 20°C), it is also correlated with the 'total density' (above the critical point this is the density of the fluid state). With the hydrothermal method the filling factor is usually between 50 and 80% and the pressure between 200 and 300 bars. Under these conditions the value of the dielectric constant is between 10 and 20, which means that water under hydrothermal conditions is still to be regarded as a polar solvent (Fig. 2). Another very important point is

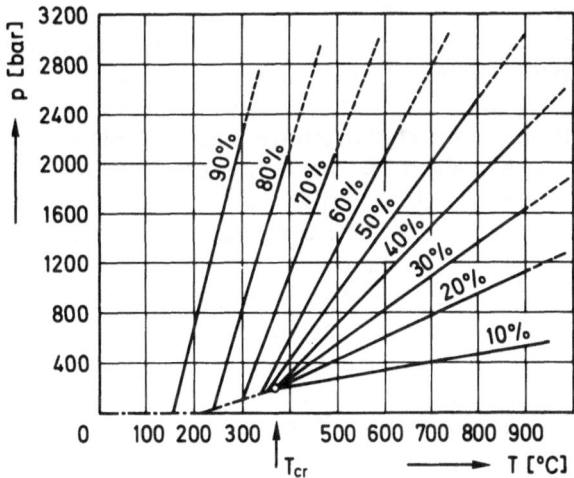

Fig. 1. Diagram relating to pure water and showing pressure as a function of temperature with the filling factor (degree of fill) of the autoclave as a parameter. The filling factor is usually between 50 and 80% and the pressure between 200 and 3000 bar. The chain-dotted line is the equilibrium line of vapor and liquid. T_{cr} is the critical temperature.

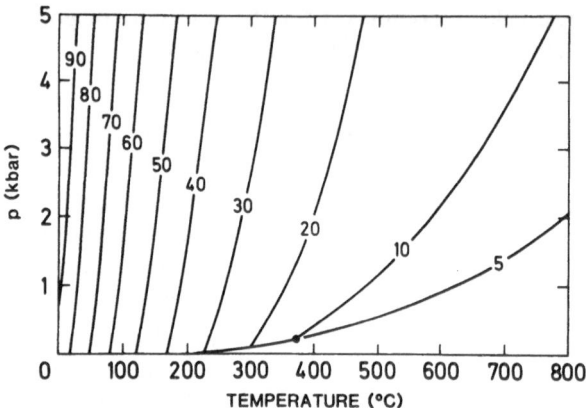

FIG. 2. Variation of the dielectric constant of water with temperature and pressure. The point • is the critical point of water.

that the ionization constant of water (K_w) increases with rising temperature and density, such that at 600°C and 2000 bar it is about 10^5 times greater than at room temperature. These conditions therefore give rise to hydrolytic reactions.

Many substances that are not soluble in water at room temperature and atmospheric pressure will show increased solubility under hydrothermal conditions. Often, however, the solubility may still not be sufficiently high: for crystal growth and synthesis it should not in practice be less than about 1%. In cases where solubility is low, readily soluble components such as acids, bases, or other complex-forming substances are also added. These substances, known in geology as 'mineralizers' form more-soluble complexes than does pure water. The solubility constant is in general temperature-dependent. This dependence is made use of by placing the material in a temperature gradient, so that the reaction product is transported from places with a high solubility to places with a low solubility. This results in the growth of relatively large crystals. This will be illustrated with an example in Section 2 of the Appendix.

Viscosity decreases with temperature; at 500°C and 100 bar the viscosity of water is 10% of its value under standard conditions. Hence, the mobility of molecules and ions in the supercritical range is much higher than under standard conditions. As a result of the low viscosity of the hydrothermal solution and of the strong convection due to the temperature gradient, the dissolved substance is transported rapidly and consequently the reaction and growth rates are generally high.

3. HISTORICAL DEVELOPMENT

Although hydrothermal synthesis remained in the domain of geological scientists until the beginning of World War II, the first description of a glass reactor comes from the German chemist Robert Bunsen (1848). He described the use of thick-walled glass (or barometer) tubes *so that in liquids an easily measurable pressure of 100–150 atmospheres can be generated without any danger.* In this way, through cooling an ammonia solution from 200°C and 15 bar, he obtained millimeter-long crystalline needles of $BaCO_3$ and $SrCO_3$. This experiment was the forerunner of visual hydrothermal synthesis, which has acquired still greater significance today. The introduction of modern hydrothermal synthesis into the geological sciences is ascribed to de Sénarmont (1851). De Sénarmont used sealed glass ampoules as reaction vessels (Fig. 3), placing them in autoclaves to avoid explo-

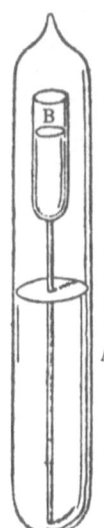

Je mettais en présence les divers agents chimiques dans des tubes de verre à demi remplis d'eau et scellés à la lampe, après qu'on y avait fait le vide. Si ces agents sont de nature à se décomposer immédiatement, on les place d'abord dans des tubes séparés A et B, et un retournement les mélange en temps opportun; on peut aussi enfermer l'une des dissolutions dans une ampoule très-mince avec une bulle d'air: la dilatation de cet air brise l'ampoule quand la chaleur est devenue assez forte.

FIG. 3. Excerpt from the publication of de Sénarmont. The ampoule B prevents the solutions from mixing with each other before A is sealed off.

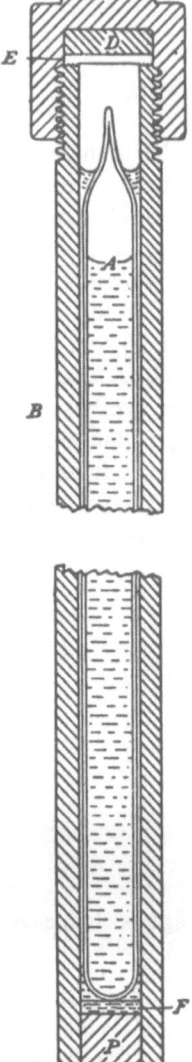

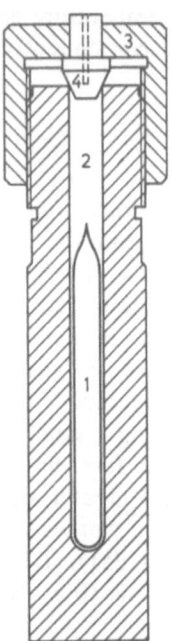

FIG. 5. Equipment for the hydro-thermal method, using hydrohalic acids as solvents. The sealed quartz ampoule (1) is filled with the material to be converted and the solvent. A calculated quantity of solid CO_2 is added to the remaining volume (2) of the autoclave so as to build up a pressure high enough to prevent the ampoule from exploding owing to the internal pressure. 3 Screw cap; 4 sealing cone with hole for thermocouple.

FIG. 4. Bomb used for heating glass tubes (Allen *et al.*, 1912).

sion. His autoclaves were fashioned from gun barrels, partially filled with water, which could be heated to a dull red glow after being welded. In spite of numerous breakdowns, he synthesized in this way a great number of oxide, carbonate, fluoride, sulfate and sulfide minerals. Among these was proustite, Ag_3AsS_3, which because of its electronic properties plays a role in modern solid-state physics. An improved version of de Sénarmont's reactor, employing Pyrex glass but again water as pressure transmitter, was reported by Allen *et al.* (1912, Fig. 4). A bibliography up to World War II, including experiments using glass ampoules, is given by Morey and Ingerson [5].

A simple technique which can easily be applied also in small laboratories without sophisticated workshops has been developed independently by Rau and Rabenau (1967, Fig. 5). This will be

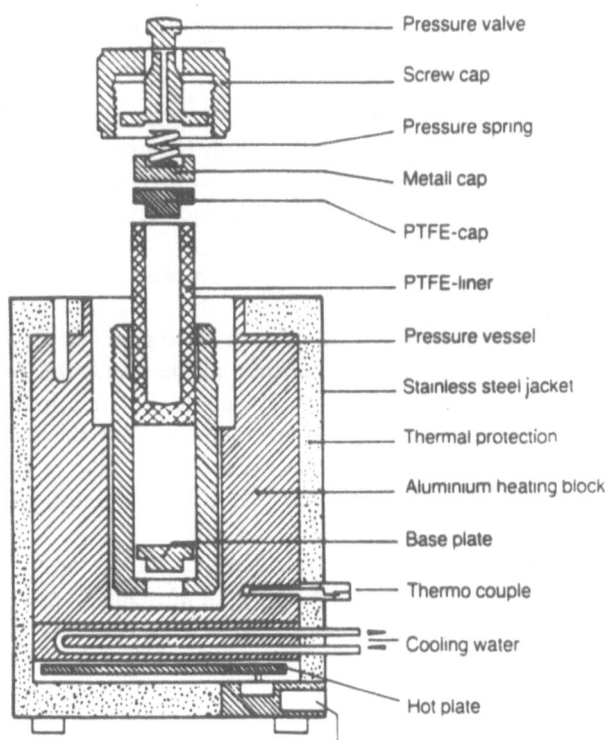

Pressure valve

Screw cap

Pressure spring

Metall cap

PTFE-cap

PTFE-liner

Pressure vessel

Stainless steel jacket

Thermal protection

Aluminium heating block

Base plate

Thermo couple

Cooling water

Hot plate

Heating connection with excess temperature safety device

FIG. 6. Digestion bomb with Teflon (PTFE) insert.

described in detail in the Appendix. From the standpoint of behavior towards corrosion, Teflon is the ideal container material. Besides its resistance to hydrofluoric acid, Teflon is also stable to alkaline media, in contrast to glass and quartz glass. Unfortunately, Teflon is only usable within a limited pressure–temperature range, since above 150°C creep behaviour can become a problem. Since it tends to be porous, only isostatically treated, potentially pore-free material (e.g. PTFE) can be used. Autoclaves with Teflon inserts have found numerous uses as digestion bombs and are commercially available: these can be used at 200°C and 200 bar (Fig. 6). There are devices, based on the same principle, capable of operating at 275°C and 350 bar. Temperatures of 300–350°C can be reached if Teflon is used as a floating insert.

4. VISUAL HYDROTHERMAL SYNTHESIS

The first example of visual hydrothermal synthesis was given by Bunsen, as described in the previous Section. He used an arrangement developed for the measurement of the tension of condensed glasses (Fig. 7), where the glass tubes had a wall thickness of 1·8 mm and a diameter of 11·2 mm.

For visual hydrothermal synthesis, safety precautions must be taken, such as working at sufficient levels below the bursting pressure and using a safety shield of 'bulletproof' glass or Plexiglass. Of the few uses of visual hydrothermal syntheses described in the literature, the works of Katsurei and Mikade, and Katsurei should be mentioned. They used ampoules of Pyrex with 10 mm diameter for thermochromic investigations up to 250°C. A quartz glass reactor with a volume of 200 ml has been described by a group of Russian investigators who used it for crystal growth up to 300°C (Fig. 8).

A suitable apparatus for the control of crystal growth has been developed for the observation of processes in ampoules under definite temperature conditions (Fig. 9).

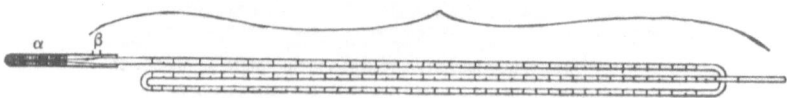

FIG. 7. Barometer tube as used by Bunsen (1839).

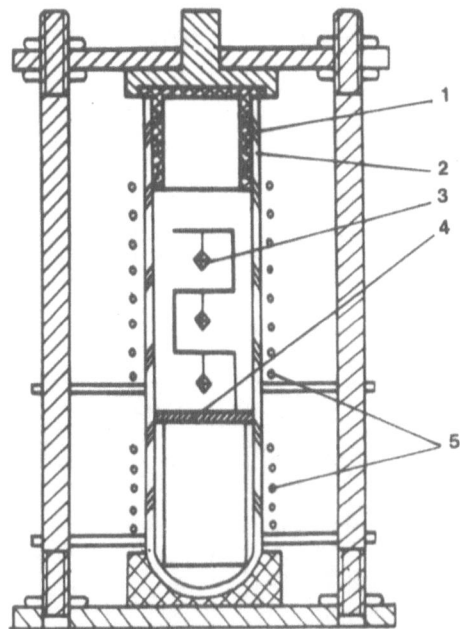

FIG. 8. Scheme of a quartz reactor for visual observation of the hydrothermal crystal growth process: 1, corrugated insert cap (made of Teflon); 2, autoclave made of fused quartz; 3, seeds; 4, baffles; 5, heater.

Quartz glass ampoules for hydrothermal investigations have also been described by Speed and Filice (Fig. 10). With the dimensions reported by the authors, the experimental bursting point at 500°C lying above 300 bar, the ampoules should be suitable for visual hydrothermal synthesis. The pressure characteristics for sealed quartz glass ampoules up to 1000°C have been reported by Holland.

5. NON-AQUEOUS SOLVENTS

Besides water—the most important solvatothermal reaction medium for transformations above the boiling point and 1 bar—there are a great number of non-aqueous solvents which, in principle, also come into consideration. As much as a century ago, Hannay and Hogarth observed that an alcoholic solution of cobalt chloride retains its characteristic blue color above the critical point and thus they

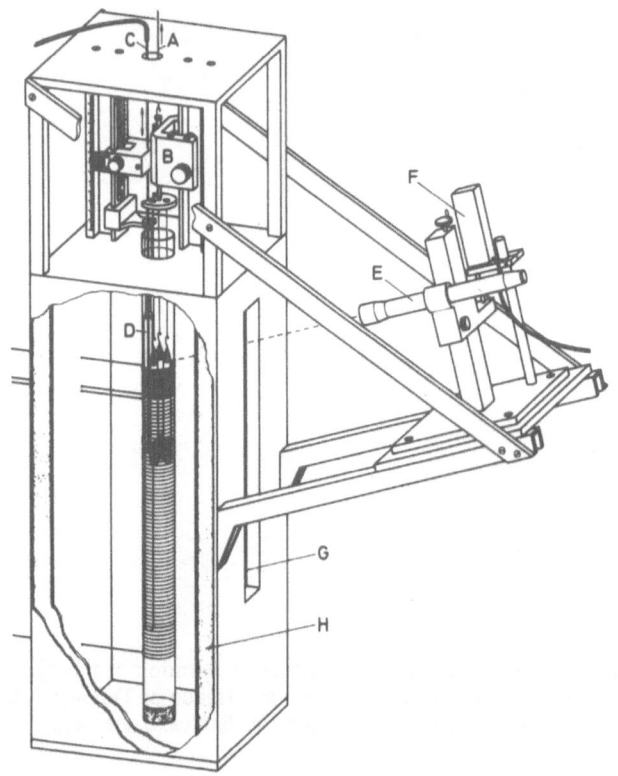

FIG. 9. Schematic sketch of the equipment used for the measurement of crystal lengths: A, wire for continuous displacement of the ampoules; B, mechanical adjustment for the ampoules; C, thermocouple; D, tube for measuring the temperature; E, telescope; F, opto-electronic length meter; G, window; H, ceramic fiber insulation.

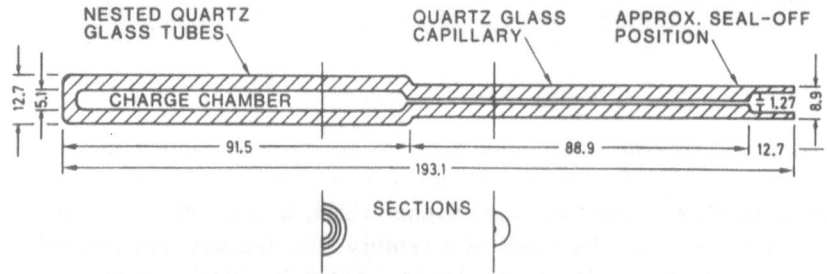

FIG. 10. Quartz glass tube after Speed and Filice. Dimensions in millimeters. The charge chamber is made by nesting three close-fitting quartz glass tubes of 1·25 mm wall thickness and fusing the tubes together in vacuum.

TABLE 1
Critical data for selected solvents

Solvent	Critical temperature (°C)	Critical pressure (bar)
Water	374·1	221·2
Ammonia	132·3	111
Chlorine	144	77·1
Hydrogen chloride	51·4	83·2
Carbon dioxide	31·3	73
Sulfur dioxide	157·8	78·7
Carbon disulfide	279	79
Hydrogen sulfide	100·4	90·1
Ethanol	243	63·8
Methylamine	156·9	40·7
Methanol	240	81
Formic acid	308	—

demonstrated the solubility of inorganic substances in a fluid medium of a non-aqueous system. Although non-aqueous solvents (with the exception of ammonia) have until now attained little significance for solvatothermal transformations, they can be considered in cases in which the required solubility cannot be reached in an aqueous medium, in cases in which the solvent itself participates in the synthesis and, most importantly, in cases where the reaction product reacts with water. Critical temperatures and pressures of non-aqueous solvents are often lower than those of water (Table 1) and the use of glass or quartz glass as reaction vessels is especially suitable.

6. CONCLUSIONS

Hydrothermal synthesis in acid or neutral solutions using glass and quartz glass as reaction vessels can be traced back more than 150 years. Well over 100 compounds, mostly minerals, have been synthesized in this way. For material synthesis, this method is still in its infancy. Extended prospects can be expected for the future by the use of non-aqueous solvents.

REFERENCES

1. A. Rabenau and H. Rau, Crystal growth and chemical synthesis under hydrothermal conditions, *Philips Tech. Rev.*, **30**, 89–96 (1969).

2. L. N. Demianets and A. N. Lobachev, Current state of the art of hydrothermal crystal synthesis, *Curr. Top. Mater. Sci.*, **7**, 483–586 (1981).
3. A. Rabenau, Methods for the study of hydrothermal crystallization, *Phys. Chem. Earth*, **13/14**, 361–74 (1981).
4. A. Rabenau, The role of hydrothermal synthesis in preparative chemistry, *Angew. Chem. Int. Ed. Engl.*, **24**, 1026–40 (1985).
5. G. W. Morey and E. Ingerson, The pneumatolitic and hydrothermal alteration and synthesis of silicates, *Econ. Geol.*, **32**, 607–761 (1937).
6. A. Rabenau and H. Rau, Chalcogenide halides of copper, gold, mercury, antimony, and bismuth, *Inorg. Synth.*, **14**, 160–73 (1973).

APPENDIX

1. Hydrothermal Synthesis in Acid Solutions: General Procedure [6]
Example: growth of antimony thioiodide crystals $Sb_2S_3 + 2\,HI = 2\,SbSI + H_2S$. The volume of the quartz glass ampoule which serves as a reaction vessel (Fig. 11) is measured up to the constriction using a graduated burette. After deducting the volume of starting material, the degree of fill is calculated from this volume. For example: volume of the ampoule 11·85 ml; volume of 10 g Sb_2S_3 about 2·15 ml; free volume 9·70 ml; 65% of 9·70 ml = 6·3 ml 10-molar HI, which is poured into the ampoule and then dipped into liquid nitrogen almost up to the constriction to freeze the acid. The powdered starting materials, here 10 g Sb_2S_3, are loaded into the ampoule with the help of a suitable funnel. After cleaning the constriction on the inside with a cotton plug, the ampoule is connected to a vacuum system, evacuated and sealed off. During this operation the level of the liquid nitrogen should be corrected from time to time to avoid thawing of the acid, which would otherwise immediately react with the starting materials. The sealed ampoule is transferred immediately from the liquid nitrogen into running hot water until a film of liquid acid has formed at the inner wall of the ampoule. In this way bursting of the ampoule by the increase in volume of the melting acid is avoided. The ampoule is then kept in a hood to allow slow thawing of the rest of the acid.

The ampoule is inserted into an autoclave (Fig. 12) and the free volume remaining in the bore hole is calibrated using water from a burette, the water being afterwards removed. Carbon dioxide is used to counterbalance the pressure inside the ampoule developed under experimental conditions. The amount of CO_2 required is read off from the graph (Fig. 13), taking into account the free volume in the

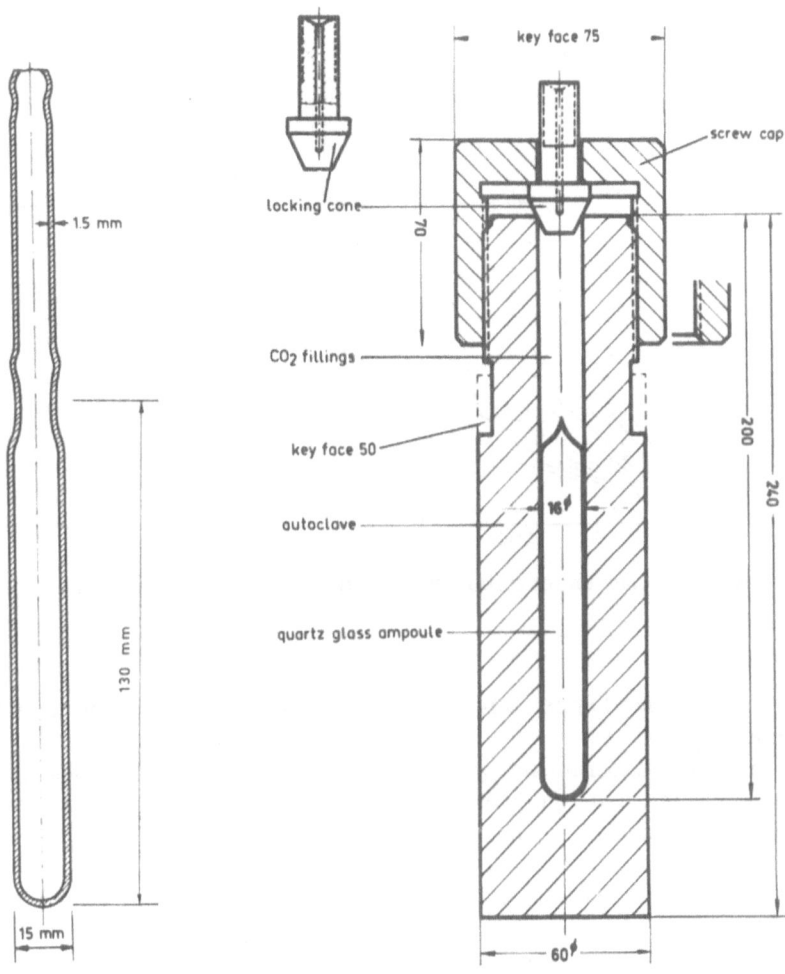

FIG. 11. Quartz glass ampoule.

FIG. 12. Autoclave; material 24 CrMoV 55: ca. 0·45 Mn, 1·35 Cr, 0·55 Mo, 0·20 (for example, CV 120 of the Deutsche Edelstahlwerke). Figure dimensions are in mm.

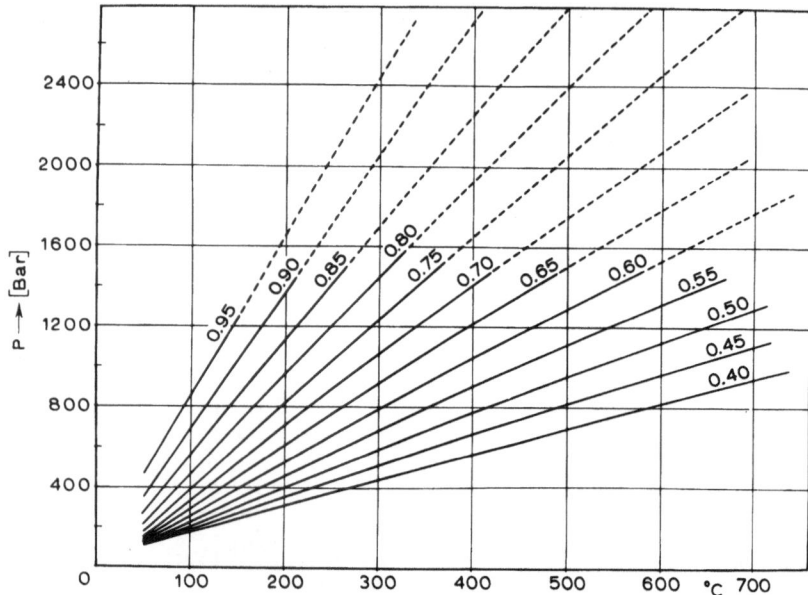

FIG. 13. $p-T$ diagram for carbon dioxide after Kennedy. Parameters in grams
CO_2 per cm^3.

autoclave, the temperature and the pressure of the experimental
conditions.

Since the experimental pressure is not known in most cases and
ampoules of the design given can support very high outside pressures,
for safety reasons, a load of 0·75 to 0·8 g CO_2 per ml of the free
volume of the autoclave should be used. A rod of dry ice (solid carbon
dioxide) of suitable diameter is prepared by placing a piece of dry ice
of sufficient size in the mould shown in Fig. 14, which is then
compressed between the jaws of a vice. A piece is cut off the rod with
a knife, the weight of the dry ice being a little more than that
calculated (plus 0·5 g to counterbalance evaporation losses during the
closing procedure). The rod is kept on a balance until the correct
weight is obtained. The piece is transferred immediately into the
autoclave above the ampoule. The locking cone is mounted and the
screw cap is bolted. Two elongated wrenches (about 1 m) are used for
this purpose. The forced applied should be sufficient to deform the
locking cone a little in the sealing region in order to make the

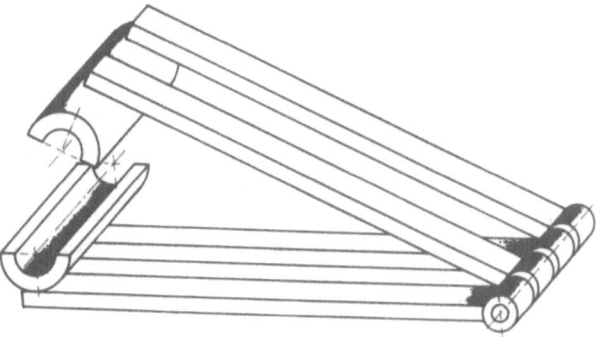

FIG. 14. Mold for pressing rods from dry ice.

autoclave gas-tight. Protective glasses must be worn during all manipulations.

The autoclave is mounted in a tube furnace with two independent filament windings to provide two temperature zones. The axis of the furnace should be at an angle of about 15° to the horizontal. This causes high convection when a temperature gradient is applied. The furnace is subjected to a temperature–time treatment. For the preparation of SbSI the furnace is heated to 490°C and cooled uniformly to 250°C over a period of 10 days. If a temperature gradient is applied, the lower part (charge zone) is at the higher temperature. Temperatures are measured with thermoelements at the ends of the autoclave. After the heating period the autoclave is allowed to cool to room temperature. The screw cap is loosened carefully until carbon dioxide escapes.

Caution: high pressure may exist inside the ampoule (H_2S, in some cases H_2). The following operations should be done in a hood behind a protective shield, for example of Plexiglas. Heavy, protective gloves should be worn. The cap is removed and the ampoule is transferred into liquid nitrogen in a Dewar vessel. The cooled ampoule is rolled in several layers of filter paper and broken open by tapping with a hammer. The contents of the ampoule are transferred to a porcelain dish and washed several times with methanol. The crystals are dried between layers of filter paper. The crystals of the respective compounds can be isolated mechanically from the accompanying material, e.g. quartz glass chips or other solid phases. The autoclave is ready for further experiments without any machining.

2. The Growth of Gold Crystals

We shall deal below in more detail with the growing of gold crystals. This brings us to a problem that has long preoccupied geologists. They have assumed that the greater part of the primary gold strata originated under hydrothermal conditions. In the laboratory, however, gold has previously been used only as a material for coating the autoclave to protect it against alkaline corrosion. This indicates that it is unlikely that the gold strata originated under such conditions.

At an earlier stage of our experiments we had already noticed that HI is an excellent solvent for gold. When 10 g of gold, in the form of wire or foil, are kept in 10-molar HI at a temperature of 500°C (with no temperature gradient), in 10 days a group of gold crystals is formed. If a temperature gradient is applied, for example, from 480 to 500°C, the gold is transported from cold to hot regions. In our further investigations we tried to find an answer to the following two questions:

(1) Can gold transport of this kind also take place under conditions actually found in nature?
(2) Is transport from cold to hot regions a reasonable assumption for these conditions?

The second question is important because all previous discussions and experiments on this subject have started from the assumption that gold is a deposit formed in the course of time from a gradually cooling solution. It has been found that gold is also transported in solutions containing hydrogen chloride and bromide, provided an oxidizing agent is present, such as free halogen or oxygen. Where the solvent is HI, the free halogen is formed by dissociation of the solvent itself.

A current view of geologists is that natural gold deposits must have been deposited at temperatures between 50 and 550°C and at pressures up to 2000 bars. In natural conditions, weakly acid salt solutions are available as solvent, with rock salt as the main constituent, in concentrations greater than 2 molar. It has been established by experiment that gold is in fact transported in weakly acid 10-molar NaCl solutions. The oxidizing agent in this case was oxygen, added in the form of H_2O_2.

We now assume that the transport of gold is based on the following equation:

$$Au(solid) + 1 \cdot 5Cl_2(gas) + Cl^-(aq) = AuCl_4^-(aq)$$

$AuCl_4^-$ is a fairly stable complex and its thermodynamic data are

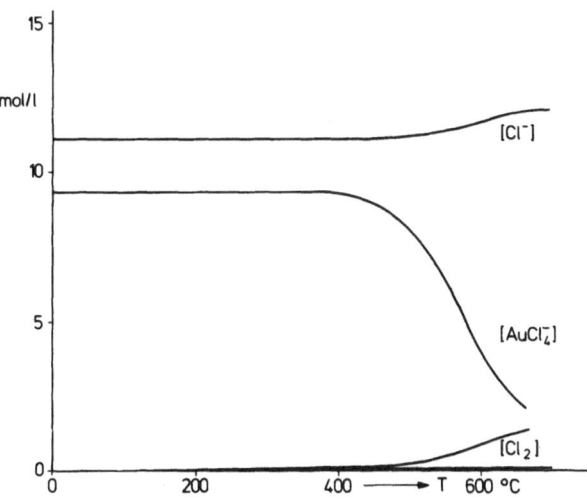

Fig. 15. Calculated concentration–temperature diagram, based on a solution equilibrium which we assume to occur in the hydrothermal growth of gold. The complex ion $AuCl_4^-$ is significant here. Gold transport from cold to hot could take place in the temperature range in which the concentration of $AuCl_4^-$ shows marked variations. (For clarity the concentration of $AuCl_4^-$ is shown on a scale 10 times larger than the calculated value).

known. These data relate to standard conditions. The values applicable to hydrothermal conditions, obtained by extrapolation, must of course be regarded as a very rough approximation. Figure 15 shows the result of a calculation based on extrapolated values. At low temperatures there is virtually no free halogen present, but as the temperature rises the complex dissociates. In the region in which the concentration of the complex varies strongly as a function of temperature we may expect a transport of gold from cold to hot. Considering the approximations used, the agreement between experiment and calculation may be regarded as satisfactory.

10

Hydrothermal Carbon: A Review from Carbon in Herkimer 'Diamonds' to that in Real Diamonds

R. C. DeVries (*Retired*)

17, Van Vorst Drive, Burnt Hills, New York, USA

ABSTRACT

On the occasion of Professor S. Sōmiya's career change, it is appropriate to consider the formation of carbon (whether amorphous, as graphite, or as diamond) by processes that seem to be basically hydrothermal. Sōmiya's commitment to this mode of synthesis and processing for oxides has resulted in many useful contributions for which he has been duly recognized. When we met, we often talked of a common interest about which neither of us did anything, namely hydrothermal carbon and particularly diamond formation by this mode. This is not a new subject, and over the past 25 years from laboratories all over the world there has been a consistent convergence on the reality of precipitation of carbon from the system C—H—O—Si. This chapter attempts to review the evidence, from amorphous carbon (anthraxolite) forming simultaneously with quartz crystals (such as Herkimer 'diamonds') under obviously hydrothermal conditions to the formation of diamond itself at depth in the earth. The next few years should bring considerable clarification and acceptance of the concept of diamond synthesis under most natural conditions, but it is unlikely that this knowledge will change the present metal–graphite mode of commercial production of synthesized diamond. Better understanding of the solubility, transport, and structure of carbon in oxides and silicates is needed.

1. INTRODUCTION

Professor Sōmiya, thank you very much for your kind invitation to participate in this lecture meeting on the occasion of your career change. My wife and I and your friends at General Electric Company in Schenectady, NY, USA wish you a long, healthy and happy new life. You have made important and well-recognized contributions to Japanese–American relationships in general and to applications of hydrothermal techniques in particular. I thank you for your friendly and sensitive hospitality in your country.

In your invitation you ask for a look into the future, with a relation to the past and present, for the particular interest of each speaker. I would like to do this for the general theme of what I'd call 'hydrothermal carbon' for lack of a better term. You and I have talked in terms of 'hydrothermal diamond' as part of our mutual concern for applications of hydrothermal processes beyond the growth of quartz and CrO_2 and a few other oxides. Because of your interest in this type of synthesis and my long puzzlement about the genesis of natural diamond in the depths of the earth, I thought it appropriate to try to bring together what has been done, is being done, and what it would be fun to do in the near future. This is not a new subject; there is a vast literature, and I draw almost totally on it to see what convergences might exist from both nature's and human laboratories and to speculate on what directions might be pursued. It is hoped that this presentation will indeed be accepted in the sense that it serve as a preliminary study for some experimentation on the subject of solution and transport of carbon by hydrothermal means. We start with some facts about carbon in oxides and silicates.

2. CARBON–OXIDE/SILICATE COEXISTENCE AND IMMISCIBILITY

Carbon exists as a mixed oxide in carbonates and oxycarbides and as carbides with many elements; but in general in the solid state, carbon coexists with oxides/silicates in a seemingly immiscible mode. Obvious examples are oil, coal and hydrocarbons in rocks, siliceous deposits in plants (both living and dead), and free carbon—as graphite veins or nodules in rocks and meteorites or as diamond in kimberlites. There is no simple solid solution between CO_2 and SiO_2 analogous to that of the latter with GeO_2, for example. (In a sense one might call a silicone a form of miscibility of Si and C in the system C–H–O–Si.) Here we consider the extent of the immiscibility and the role of the hydrothermal environment in the solution and precipitation of carbon in oxide systems. There is a large body of knowledge pertinent to the system C–H–O–Si that would seem to be a common basis for understanding. We begin with some selected examples of carbon–oxide/silicate occurrences from synthesized materials and then consider natural systems.

2.1. Synthesized Materials

(1) Carbon impurity in synthetic quartz is reported at concentrations two orders of magnitude higher than the usually detected impurities.

(2) Carbon is reported as an impurity in synthetic arc-fusion crystals of MgO and Mg_2SiO_4 to the extent of a few tens to a few hundreds of parts per million. The form is considered to be atomic C (Freund *et al.*, 1977, 1980).

(3) Solution of carbon and reprecipitation as graphite crystals was observed in high-pressure experiments with CdO, Cu_2O, ZnO, pyrophyllite, and some molten silicate systems containing OH and K and Na. Some halide systems were also found to dissolve carbon. Carbon is considered to be able to be present as neutral atoms and as both positively and negatively charged species. The solubility range was up to about 1 wt% (Wentorf, 1966).

(4) Carbon was dissolved in molten $3CaO \cdot Al_2O_3$ at high temperatures to form a clear glass and could then be precipitated as graphite crystals by annealing (Nestor, 1967). Solubilities to about 4 wt% were reported. This result is probably related to solubility in the $CaO–Al_2O_3–Al_4C_3$ system. The oxycarbides, Al_4O_4C and Al_2OC, are stable phases in the $Al_2O_3–Al_3C_4$ system (Foster *et al.*, 1956), and if oxycarbide formation is a form of solubility, then obviously rather large amounts of carbon can be present. Carbon contents to 0·35 wt% have been found for Mg–Al–Si–O–C glasses (Homeny *et al.*, 1988).

(5) About 7 and 2 wt% carbon can be dissolved in basic and acidic silicate liquids, respectively, as determined by analysis of glasses quenched from 30 kbar and 1625°C (Eggler, 1978). However, crystalline phases from similar conditions contain only 7–10 ppm C.

(6) There are patented processes for precipitation of both diamond and graphite by the reaction of oxides with carbonates at high pressures and temperatures—in other words, the separation of elemental carbon by reduction of carbonates (Woermann *et al.*, 1981).

Summary. With the exception of OH in materials in case (3) above, evidence exists for a small but finite solubility of carbon by oxygen-containing systems under essentially dry or non-hydrothermal conditions. The disordered state can dissolve more than the crystalline state for silicates.

2.2. Natural Systems

(1) Cancrinite is a mixed silicate–carbonate mineral from nepheline syenites ($3NaAl_2Si_2O_8 \cdot 2CaCO_3$). This phase has also been made synthetically. It behaves as both a carbonate and a silicate in chemical reactions with acids. There is also a report of mixed oxide carbonate based on Mg.

(2) Hydrocarbons are found in stony meteorites and silicates as inclusions disseminated in the graphite phase of iron meteorites.

(3) Silicate phases such as garnet, enstatite and other phases are found as inclusions in diamond in kimberlites and lamproites and are the basis for establishing the $P-T$ conditions for the growth of diamond at depth in the earth (Boyd *et al.*, 1985; Cohen and Rosanfeld, 1979; Hervig *et al.*, 1980; Meyer & Tsai, 1976).

(4) Carbon inclusions are found in silicates such as olivine, serpentinized olivine, garnet and andalusite. The chiastolite variety of andalusite is characterized by carbonaceous inclusions zoned along certain directions of the crystal, presumably by rejection as the crystal grew. Hydrocarbon inclusions are found in garnets in kimberlites, and diamond crystals are found in garnets from reactions rims between a granite and a graphite gneiss (Dergachev, 1986; Garanin *et al.*, 1988). This chemistry is quite different from that of basic magma normally associated with diamond.

(5) There are many examples of graphite veins and crystals in silicate and carbonate rocks; some of these have clearly been given a hydrothermal origin (Rumble *et al.*, 1986).

(6) Diamond and graphite in kimberlite and lamproites and eclogites are considered by most to be xenoliths in these rocks. The mystery of this occurrence is the driving force for most of the research on carbon in silicates. Diamonds in kimberlites always show evidence of high-temperature etching in the form of trigons.

(7) Anthraxolite occurs in quartz crystals (Herkimer diamonds). This carbon–SiO_2 coexistence is discussed in more detail later.

(8) The solubility of quartz in natural water is reported to be increased by the presence of dissolved organic complexes from the microbial degradation of crude petroleum (Bennett and Siegel, 1989).

Summary. Carbon and oxide/silicates can coexist in a wide variety of environments in nature; in many occurrences there is also conclusive evidence for hydrothermal conditions. The solubility and transport of carbon is certainly more extensive in nature than in the synthetic examples (with the possible exception of carbon in aluminate glass), and perhaps hydrothermal conditions have something to do with this. We now consider the two extremes of $P-T$ conditions represented by the title of this chapter: Herkimer 'diamonds' (quartz crystals) and real diamonds.

2.2.1. Anthraxolite and quartz crystals (Herkimer Diamonds)

These crystals are found in solution cavities or vugs in siliceous dolomites near Herkimer, New York State, and in many other locations in the world. These attractive crystals are commonly double-ended pyramids, as if they have had no attachment site (in common with most real diamonds). They have been supported in some way during growth. They are included in and often enclose within themselves a black, carbonaceous (>90% C, poorly crystallized) material called anthraxolite (Dunn and Fisher, 1954). In Russia a similar material is called shungite. Sometimes rather rounded quartz crystals are found totally embedded in the anthraxolite. The quartz crystals contain cavities with both liquid and gas and therefore clearly were formed under hydrothermal conditions which were not extreme enough to crystallize the carbon into graphite. The conclusion from the only detailed study of this deposit is that the carbon came from carbonaceous material previously disseminated in the dolomite and not from decomposition of the carbonate. However, the coexistence of considerable amounts of SiO_2 along with a carbon-rich phase during growth of quartz crystals is all we want to note from this occurrence. It probably represents a mild $P-T$ condition where perhaps organic complexes influenced the solubility of the SiO_2. It seems reasonable to extrapolate from this kind of a deposit to deeper-seated igneous and metamorphic environments where higher pressures and temperatures would yield graphite and diamond instead of anthraxolite. Immiscibility of a carbon-rich liquid and a silicate liquid would be common to both environments.

2.2.2. Graphite and diamond in silicates

Although the origin of natural diamonds is not clearly defined and still invokes controversy, there is considerable convergence on a deep-seated origin involving the system C–H–O–Si. There certainly is very little evidence that the synthetic diamond systems involving catalysis or solution/catalysis with group VIII metals plus a few others were operative. Other than the presence of some iron sulphides and oxides in natural diamond, there is no undisputed evidence for primary native metal included in diamond from deep-seated deposits. When synthesizing diamonds from metal–carbon systems, the presence of H–O–C is deleterious; but it is possible to synthesize well-formed diamond directly from hydrocarbons without the presence of a metal solvent–catalyst (Serebryanaya *et al.*, 1985).

On the other hand there are well-established data on the presence of silicate inclusions and of C, H, O, and N in diamonds from kimberlites and lamproites. C–O, C–H, and O–H gas species are found when diamonds are fractured under vacuum and analysed in a mass spectrometer. It has been stated that H is as important an impurity in diamond as N. Similar species are found for synthetic MgO and for both natural and synthetic olivine, and olivine has been suggested as a water-containing phase at depth in the mantle. There is clear evidence for hydrothermal graphite in vents and in metamorphic rocks. Of particular interest are the reports of diamond and graphite growth in a reaction rim between a granite and a granite–gneiss during the process of boudinage; these are not basic rocks like kimberlite but acid rocks often associated with hydrothermal activity. The implication is that the pressure–temperature conditions were within the diamond stability range because of stresses in the shear zone between the intruding mass and the host, and in addition that there was a means of carbon transport (Garanin *et al.*, 1988).

Between these two extremes of an uncrystallized carbon such as anthraxolite with quartz and well-crystallized diamond (and graphite) in kimberlites and other deep-seated deposits, there are many graphite deposits from intermediate metamorphic rocks that are clearly associated with hydrothermal conditions. Hydrothermal graphite in which the carbon was mobilized from sediments as CO_2 and CH_4 during metamorphic devolatilization reactions and transported through fractures in aqueous fluids has been described in a New Hampshire deposit (Rumble *et al.*, 1986). Graphite crystals are found in hydrothermal vents on the ocean floor along with antimonide overgrowths (Jedwab and Boulegue, 1984).

3. THE SYSTEM C–H–O–Si

Ultimately the detailed explanation of the carbon chemistry in the earth will be in terms of this system. This has been recognized for a long time, and the pioneers of hydrothermal experimental work, O. F. Tuttle, Wyllie, Eggler, and others, contributed to this knowledge. The early work seemed biased toward the phase relations of carbonate or carbonatites and silicates rather than carbon *per se* and silicates. But clearly the accumulation of experimental data, especially in the last 10 years, plus the theoretical work of French (1966), Langford (1978), Frost (1979) and others have converged on the coexistence of a liquid phase containing various species such as CO, CO_2, CH_4, H_2, H_2O with silicates. The composition tetrahedron (Fig. 1) is useful as an overall perspective on the reactions listed in Table 1. However the most important information is now being summarized by Haggerty (1986), by Ryabchikov *et al.* (1985), and by others in the form of diagrams which show phase stability regions in terms of pressure (depth), temperature, and oxygen fugacity. Such a representation by Haggerty is shown in Fig. 2. There is a vast literature accumulating on this subject, of which only about 200 papers were studied for this review. In a manner which is unfair to this body of information for the purposes of this review, we simply accept the general agreement that C–H–O fluids in this system are the basis for the inorganic solution, precipitation and growth of graphite and diamond crystals from the surface to the mantle. The carbon fluids are considered to be essentially immiscible with the host silicate liquid or rock. The amount of carbon liquid may be quite small and limited to grain boundaries or small pockets. Whether diamond or graphite occurs is primarily a matter of which stability region was reached, although there might be some solvent influence. Any supply of carbon at depth greater than about 150 km in the crust has a high probability of becoming diamond, although from studies in metal solutions it is well known that graphite can also grow in the diamond-stable region. We suggest that there are probably several factors contributing to the dearth of larger diamonds

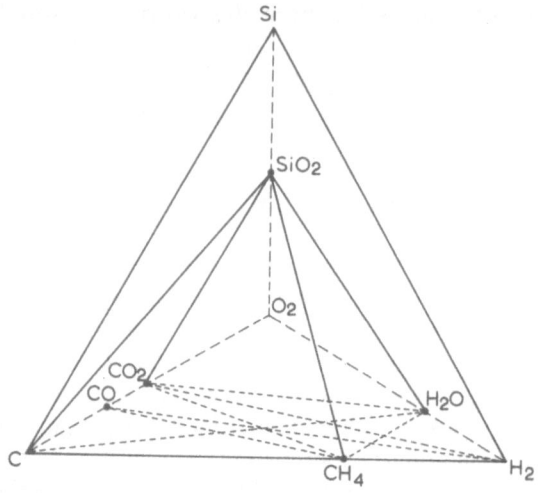

FIG. 1. Composition tetrahedron for C–H–O reactions in the C–H–O–Si system.

from such huge pressure vessels: an erratic and limited supply of carbon-rich fluids; overpressure which would lead to multiple nucleation; and, of course, supposedly a difficult trip up from depths.

There is little evidence for metastable growth of diamond in earth systems; but there is general agreement based on studies of the silicates included in diamond that it forms only above about 1100°C and at pressures greater than about 45 kbar. Although the hydrocarbon–H_2 chemistry of metastable chemical vapor deposition of diamond (currently receiving so much attention in Japan and the rest of the world) is indeed a part of the C–H–O–Si system, this mode of growth is more likely applicable to diamond nucleation and growth in outer space than to most natural diamond.

TABLE 1
C–H–O reactions

$C + CO_2 = 2CO$	(Boudouard)
$C + H_2O = CO + H_2$	(water gas)
$2C + 2H_2O = CO_2 + CH_4$	
$C + 2H_2O = CO_2 + 2H_2$	$CO_2 + 2H_2O = 2O_2 + CH_4$
$3C + 2H_2O = 2CO + CH_4$	$CO + 3H_2 = H_2O + CH_4$

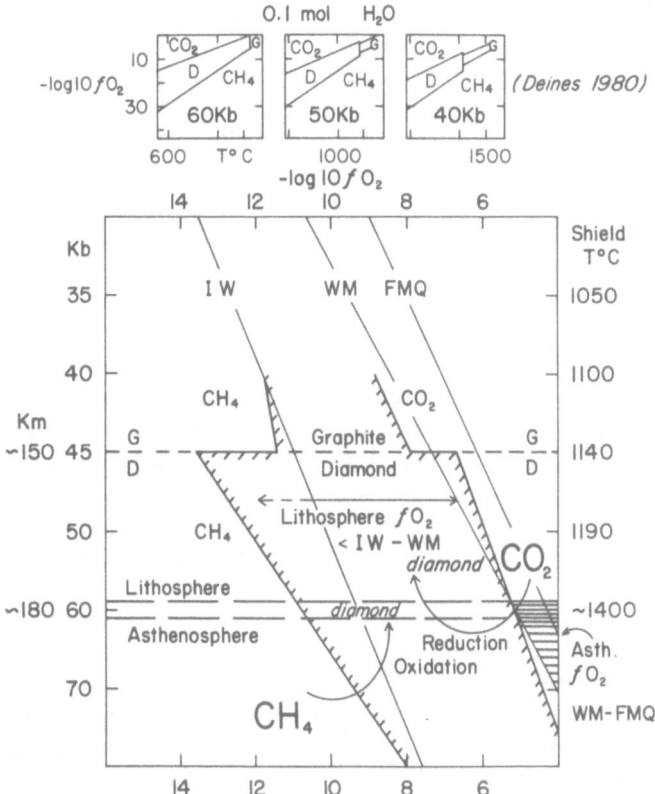

Fɪɢ. 2. Diamond–graphite relations in the system C–O–H as a function of P, T, and $f(O_2)$ in the presence of 0·1 mol H_2O (From Haggerty, 1986).

There still is controversy about the details of the chemistry of the fluids and the partitioning of species between solids and fluids. The important determination of oxygen fugacities in the complex systems will continue, but the importance of species in the C–H–O–Si system at high pressures and temperatures seems to have a general consensus.

Solubility and transport mechanisms. A real area of controversy has to do with how carbon is present in silicates and oxides, whether molten or solid, and how is it transported in these media. The work of Wentorf, Freund, Eggler, Boettcher, Mysen, etc., should be consulted. We consider here very briefly the experimental and speculative findings of these workers.

Wentorf—His work was mentioned briefly above. He was able to grow graphite, but not diamond, at high pressures and temperatures from solution of carbon in CdO, ZnO, Cu_2O, pyrophyllite, and from some molten silicates containing OH, and Na and K plus OH as serpentine and mica. He found similar results with some halides and sulphides. Solubility of carbon was on the order of 1 wt%, but was negligible in $CaCO_3$ at 60 kbar and 1300°C. From his studies of transport of carbon in various media using d.c. fields, he concluded that carbon could have a net positive (in Fe or Ni), zero (as in oxides), or net negative charge (as in CaC_2 or LiC_2). Diamond is favoured from solutions in which carbon is positive in the diamond stability region. Wentorf suggests a possible explanation for the influence of the net charge on precipitation of diamond or graphite based on the decreasing size of the atom with loss of electrons. However, at very high pressures, e.g. above the direct transistion pressure for graphite to diamond (120–150 kbar) nucleation of diamond may predominate no matter what the solvent. There is still a need for more experimentation on the solution of carbon in various solvents at high pressures and temperatures.

Eggler, Boettcher, Mysen—A large amount of phase equilibria data on silicate–H_2O–CO_2 systems exists showing the sometimes spectacular lowering of melting points of silicates and carbonates by solution of H_2O and CO_2. More recent emphasis is on the mechanism of solution and on spectrometric studies of the structure of solutions. According to Boettcher, carbon dissolves as CO_2, as CO_3^{2-}, or possibly as tetrahedral carbon in aluminosilicate liquids. Experimental data for the last of these modes is lacking. This group suggests that the low solubility of carbon in silicate liquids compared to that of hydrogen and water reflects the low concentration of sites appropriate for carbon in any form, which is another way of describing the immiscibility of carbon. Other factors influencing the solubility of carbon are decreasing polymerization of liquids (enhanced solubility); increasing concentration or activity of SiO_2 (increases CO_2/CO_3^{2-}); availability of mono- and divalent cations for complexing CO_3^{2-} (enhanced solubility of C as CO_3^{2-}. This brief overview does no justice to this active area of research; it merely points out the questions being asked. There is considerably difficulty in obtaining structural information on the species present in the solids and liquids, particularly at pressure.

Freund—He and his colleagues started with questions about carbon in arc-fusion cast MgO and them Mg_2SiO_4 and then went on to extrapolate to oxides and silicates in the crust and mantle. His concept is fascinating and controversial, particularly for those doing the more traditional hydrothermal solubility studies. Some of the disagreement may be resolved by more sophisticated spectrometric work under difficult high-pressure–high-temperature conditions. It seems to this author that Freund looks at this problem in a unique way that raises entirely different questions about C, O, and H in oxides and silicates. The experimental proof has been elusive. Freund states, for example, that dissolved water in minerals forms not only OH^- but also O_2^{2-} and H_2 via charge-transfer processes. Excess oxygen can be retained as O with a valence of -1 if H_2 is lost by diffusion. He believes that CO_2 in non-carbonate minerals forms not CO_3^{2-} but CO_2 and O_2^{2-}. Unusual bonding features are claimed for the CO_2^{2-} permitting rapid diffusion of carbon. In this state carbon is in a reduced, nearly zero-valent state (atomic). The charge-transfer states are smaller in volume than normal states and are therefore favored by high pressure.

Summary. There is considerable agreement for fluids described in the system C–H–O–Si as a primary source and mode of transport of carbon in oxide and silicate systems. Disagreement exists on the details of how carbon is present structurally in these systems. If indeed the concept of an immiscible carbon-rich fluid proves to be an important mechanism, the concept of transport may reduce to one of grain-boundary migration rather than having to deal with diffusion through silicate structures, whether they be glassy or crystalline.

4. SUGGESTIONS FOR R & D

In the spirit of Sōmiya's crystal ball and what to do next, we consider hydrothermal carbon in terms of both understanding and possible applications.

4.1. Understanding

(1) The need for both experimental verification and theoretical support for the structure of silicate–carbonate–water solutions has already been mentioned. This is a difficult experimental problem, perhaps approachable in the diamond anvil for the higher-pressure regime, using spectroscopy to check the charge-transfer model.

(2) There is need for some hydrothermal studies on carbon itself in the graphite stable regime at about 1–2 kbar and 600°C in water alone to reveal something about the mechanism of solution and transport of carbon. We need to start with different types of carbon of varying degrees of graphitization and see whether there is any effect of small amounts of Fe on graphite formation as there in the CVD systems at about 1100°C with CH_4/H_2 decomposition. One could put in some SiO_2 and see if there are organic complexes that increase the solubility of this oxide.

(3) Enlargement of the system to include additives such as F, S, N and some of the elements and compounds that are known to intercalate with graphite. Perhaps some of these 'intercalators' will influence the kinetics of graphitization by making it easier for the planes to become ordered. Also, there are enough citings in the literature of the possible role of iron sulphides in the genesis of diamond to warrant a detailed look at this chemistry. These experiments should be done first in the graphite-stable region to lessen the experimental problems.

(4) Experiments with silicones as starting materials for studies of the system C–H–O–Si under hydrothermal conditions are warranted.

4.2. Possible Applications

(1) The graphitization of carbon under hydrothermal conditions is discussed above under 'understanding'. It would be fun to see whether this has practical significance. Nature obviously graphitized carbon at much lower temperatures than are used in dry systems, but she did not have to worry about kinetics and large pressure vessels were available. What is the effect of the hydrothermal environment on this important industrial concern? Perhaps some of the additives suggested above will help the crystal planes to become lined up more easily in a hydrothermal environment. Professor S. Hirano essentially asked this question and did some experiments with water additions in his thesis in 1970. He found some mixed effects depending on which Ca compound was present.

(2) Since carbon can coexist with oxides, it should be possible to infiltrate masses of it with oxides under hydrothermal conditions to produce some composites that cannot be made any other way because of reactivity or volatilization.

(3) Since carbon can be dissolved in calcium aluminate, it would be interesting to study the effect of its controlled precipitation on

the properties of the resulting composite. Since this material appears to be in a class by itself with respect to solubility of carbon, it would also be interesting to find out why and whether there are other such materials based on analogy with the system $CaO-Al_2O_3-Al_4C_3$.

4.3. Not Possible Applications

Unless one wants to consider the addition of OH via saké or alcohol (Hirose's method) to CH_4/H_2 mixtures for CVD diamond as a hydrothermal process, it does not seem that large-scale synthesis of diamond by hydrothermal means is practical. It will not replace the more efficient industrial approach of packing the cell with the metal–graphite mixtures. Nature's yield and quality control by whatever process she used was extremely inefficient.

5. SUMMARY

This attempt to bring together a number of observations and ideas pertinent to carbon solubility and transport by hydrothermal processes has been a superficial sweep of a complicated field. The best it may do it trigger some better questions than those raised. Perhaps Professor Sōmiya and his students may find this area intriguing enough to discover new facts and concepts about how carbon is present and is transported in oxides and silicates.

BIBLIOGRAPHY

Akimoto, S. and Akaogi, M. (1980). The magnesium silicate (Mg_2SiO_4)– magnesium oxide–water system at high pressures and temperatures— possible hydrous magnesian silicates in the mantle transition zone. *Phys. Earth Planet. Inter.*, **23**, 268–75.

Alekseevskii, K. M., Botkunov, A. I., Nikolaeva, T. T., Ermilov, V. V. and Nastasienko, E. V. (1985). Chemical changes of the environment of diamond synthesis. In *Vopr. Orudeneniya Ul'tramafitakh* ed. I. F. Romanovich. Nauka, Moscow.

Aleshin, V. G., Bogatikov, O. A., Kononova, V. A., Novgorodova, M. I., Smekhnov, A. A., Novikov, N. V. and Nemoshkalenko, V. V. (1986). Relics of reducing fluids trapped in native metals. *Dokl. Akad. Nauk SSSR*, **291**, 957–60.

Arculus, R. J., Dawson, J. B., Mitchell, R. H., Gust, D. A. and Holmes, R. D. (1984). Oxidation states of the upper mantle recorded by megacryst ilmenite in kimberlite and type A and B spinel lherzolites. *Contrib. Mineral Petrol.*, **85**, 85–94.

Bennett, P. and Siegel, D. I. (1987). Increased solubility of quartz in water due to complexing by organic compounds. *Nature*, **326**, 684–6.

Berg, G. W. (1986). Evidence for carbonate in the mantle, *Nature* **324**, 50–1.

Bezrukov, G. N. (1974). Genesis of diamond in light of experimental studies on its artificial preparation. *Sov. Geol.*, 31–40.

Bleshinskii, S. V. (1984). New trends in the study of Lower Paleozoic carbonaceous shales. *Izv. Akad. Nauk Kirg. SSR*, 38–43.

Boettcher, A. L. (1984). The system silica–water–carbon dioxide: melting, solubility mechanisms of carbon, and liquid structure to high pressures. *Am. Mineral*, **69**, 823–33.

Boettcher, A. L., Luth, R. W. and White, B. S. (1987). Carbon in silicate liquids: the systems $NaAlSi_3O_8-CO_2$, $CaAl_2Si_2O_8-CO_2$, and $KAlSi_3O_8-CO_2$. *Contrib. Mineral. Petrol.*, **97**, 297–304.

Bokii, G. B. (1982). Chemical transport of carbon by nitrogen-containing 'intermediates' in natural diamond synthesis. *Dokl. Akad. Nauk SSSR*, **266**, 711–14.

Bonijoly, M., Oberlin, M. and Oberlin, A. (1982). A possible mechanism for natural graphite formation. *Int. J. Coal Geol.*, **1**, 283–312.

Botkunov, A. I. (1984). Redox conditions during diamond growth. *Byul. NTI. Yakut. Fil. SO AN SSSR*, 22–5.

Botkunov, A. I., Garanin, V. K., Krot, A. N., Kudryavtseva, G. P. and Matsyuk, S. S. (1985). Primary hydrocarbon inclusions in garnets from the Mir and Sputnik kimberlite pipes, *Dokl. Akad. Nauk SSSR*, **280**, 468–73.

Boyd, F. R. and Finnerty, A. A. (1980). Conditions of origin of natural diamonds of peridotite affinity. *J. Geophys. Res. B*, **85**, 6911–18.

Boyd, F. R., Gurney, J. J. and Richardson, S. H. (1985). Evidence for a 150–200 km thick Archean lithosphere from diamond inclusion thermobarometry. *Nature* **315**, 387–9.

Boyd, S. R., Mattey, D. P., Pillinger, C. T., Milledge, H. J., Mendelssohn, M. and Seal, M. (1987). Multiple growth events during diamond genesis: an integrated study of carbon and nitrogen isotopes and nitrogen aggregation state in coated stones. *Earth Planet. Sci. Lett.*, **86**, 341–53.

Brey, G. P. and Green, D. H. (1976). Solubility of carbon dioxide in olivine melilitite at high pressures and role of carbon dioxide in the earth's upper mantle. *Contrib. Mineral. Petrol.*, **55**, 217–30.

Bulanova, G. P. and Pavlova, L. P. (1987). Magnesite peridotite assemblage in diamond from the Mir pipe. *Dokl. Akad. Nauk SSSR*, **295**, 1452–6.

Burton, K. W. (1986). Garnet-quartz intergrowths in graphitic pelites: the role of the fluid phase. *Mineral. Mag.*, **50**, 611–20.

Casquet, C. (1986). Carbon-oxygen-hydrogen-nitrogen fluids in quartz segregations from a major ductile shear zone: the Berzos fault, Spanish Central System. *J. Metamorph. Geol.*, **4**, 117–30.

Chaussidon, M., Albarede, F. and Sheppard, S. M. F. (1987). Sulphur isotope heterogeneity in the mantle from ion microprobe measurements of sulphide inclusions in diamonds, *Nature*, **330**, 242–4.

Chou, I. M. (1987). Calibration of the graphite-methane buffer using the f/h2 sensors at 2-kbar pressure. *Am. Mineral.*, **72**, 76–81.

Cohen, L. H. and Rosenfeld, J. L. (1979). Diamond: depth of crystallization inferred from compressed included garnet. *J. Geol.*, **87**, 333–40.

Dahe, X. and Zhaotian, L. (1985). Experimental studies of calcium carbonate as a carbon source for synthesizing diamonds. *Scientia Geologica Sinica*, 255–65.

Dergachev, D. V. (1986). Diamonds of metamorphic rocks. *Dokl. Akad. Nauk SSSR*, **291**, 189–91.

Dergachev, D. V. (1984). Diamond crystal growth in a diffuse stream of juvenile carbon. *Geol. Rud. R-nov i Mestorozhd. Tverd. Polez. Iskopaemykh Kazakhstana, Alma-Ata*, 142–51.

Dickey, J. S., Jr., Bassett, W. A., Bird, J. M. and Weathers, M. A. (1983). Liquid carbon in the lower mantle. *Geology*, **11**, 219.

Dickinson, J. T., Jensen, L. C., McKay, M. R. and Freund, F. (1986). The emission of atoms and molecules accompanying fracture of single crystal MgO, *J. Vac. Sci. Technol.*, **A4**(3), 1648–52.

Diessel, C. F., Brothers, R. N. and Black, P. M. (1978). Coalification and graphitization in high-pressure schists in New Caledonia, *Contrib. Mineral. Petrol.*, **68**, 63–78.

Draper, D. and Goodchild, W. H. (1916). Notes on the genesis of diamond. *Mining J.*, **113**, 357–9.

Draper, D. and Goodchild, W. H. (1930). Notes on the genesis of diamond. *S. African Mining and Eng. J.*, **40**, Pt. 2, 495–6, 570–2.

Dubessy, J. (1984). Simulation of chemical equilibria in the carbon–oxygen–hydrogen system. Methodological consequences for fluid inclusions. *Bull. Mineral.*, **107**, 155–68.

Dunn, J. and Fisher, D. (1954). Occurrence, properties and paragenesis of anthraxolite in the Mohawk Valley. *Am. J. Sci.*, **252**, 489–501.

Eggler, D. H. (1975). CO_2 as volatile component of the mantle: the system Mg_2SiO_4–SiO_2–H_2O–CO_2. *Phys. Chem. Earth*, **9**, 869–81.

Eggler, D. H. (1976). Does CO_2 cause partial melting in the low-velocity layer of the mantle? *Geology*, 69–72.

Eggler, D. H. (1978). Carbon dioxide in silicate melts: II Solubilities of CO_2 and H_2O in $CaMgSi_2O_6$ (diopside) liquids and vapors at pressures to 40 kb. *Am. J. Sci.*, **278**, 64–94.

Eggler, D. H., Mysen, B. O., Hoering, T. C. and Holloway, J. R. (1979). The solubility of carbon monoxide in silicate melts at high pressures and its effect on silicate phase relations. *Earth Planet. Sci. Lett.*, **43**, 321–30.

Eggler, D. H., Kushiro, I. and Holloway, J. R. (1979). Free energies of decarbonation reactions at mantle pressures: I. Stability of the assemblage forsterite–enstatite–magnesite in the system MgO–SiO_2–CO_2–H_2O to 60 kbar. *Am. Mineral.*, **64**, 288–93.

Eggler, D. H. (1983). Upper mantle oxidation state: evidence from olivine–orthopyroxene–ilmenite assemblages. *Geophys. Res. Lett.*, **10**, 365–8.

Ellis, D. E. and Wyllie, P. J. (1979). Carbonation, hydration, and melting relations in the system MgO–H_2O–CO_2 at pressures up to 100 kbar. *Am. Mineral.*, **64**, 32–40.

Ellis, D. E. and Wyllie, P. J. (1980). Phase relations and their petrological implications in the system $MgO-SiO_2-H_2O-CO_2$ at pressures up to 100 kbar. *Am. Mineral.*, **65**, 540–6.

Eyring, H. and Cagle, F. W., Jr. (1952). An examination into the origin, possible synthesis and physical properties of diamonds. *Z. Elektrochem.*, **56**, 480–3.

Fedoseev, D. V. and Deryagin, B. V. (1982). Formation of metastable forms of carbon in the gas phase. *Izv. Akad. Nauk SSSR., Ser. Khim.*, 1725–9.

Fein, J. B. and Walther, J. V. (1987). Calcite solubility in supercritical carbon dioxide–water fluids. *Geochim. Cosmochim. Acta,* **51**, 1665–73.

Fesq, H. W., Bibby, D. M., Erasmus, C. S. and Kable, E. J. D. (1975). National Institute of Metals, Johannesburg, South Africa, Report NIM-1636.

Foster, L. M., Long, G. and Hunter, M. S. (1956). Reactions between aluminum oxide and carbon; the $Al_2O_3-Al_4C_3$ diagram. *J. Am. Ceram. Soc.*, **39**, 1–11.

Frank, F. C., Lang, A. R. and Moore M. (1973). Cavitation as a mechanism for the synthesis of natural diamonds. *Nature,* **246**, 143–4.

Franz, G. W. and Wyllie, P. J. (1979). Experimental studies in $CaO-MgO-SiO_2-CO_2-H_2O$. In *Ultramafic and Related Rocks,* ed. P. J. Wyllie, Robert E. Krieger Publishing Co, Huntington, N.Y.

French, B. M. and Rosenberg, P. E. (1965). Siderite ($FeCO_3$): Thermal decomposition in equilibrium with graphite. *Science,* **147**, 1283–4.

French, B. M. (1966). Some geological implications of equilibrium between graphite and a C–H–O gas phase at high temperatures and pressures. *Rev. Geophys.,* **4**, 223–53.

Freund, F., Debras, G. and Demortier, G. (1977). Carbon content of magnesium oxide single crystals grown by arc fusion method. *J. Crystal Growth,* **38**, 277–80.

Freund, F., Debras, G. and Demortier, G. (1978). Carbon content of high-purity alkaline earth oxide single crystals grown by arc fusion. *J. Am. Ceram. Soc.,* **61**, 429–34.

Freund, F., Kathrein, H., Wengeler, H. and Knobel, R. (1980). Carbon in solid solution in forsterite—a key to the untractable nature of reduced carbon in terrestrial and cosmogenic rocks. *Geochim. Cosmochim. Acta,* **44**, 1319–33.

Freund, F. *et al.* (1980). Atomic carbon in magnesium oxide, Part I: Carbon analysis by the 12C(d,p)13 method. *Mater. Res. Bull.,* **15**, 1011–18; Part II; Laserflash-induced mass spectrometry. *Mater. Res. Bull.,* **15**, 1019–24.

Freund, F. (1981). Charge transfer and O^- formation in high and ultrahigh pressure phase transitions. *Bull. Mineral.,* **104**, 177–85.

Freund, F. (1981). Mechanism of the water and carbon dioxide solubility in oxides and silicates and the role of O^-. *Contrib. Mineral. Petrol.,* **76**, 474–82.

Freund, F. (1982). Solubility mechanisms of water in silicate melts at high pressures and temperatures: a Raman spectroscopic study: discussion. *Am. Mineral.,* **67**, 151–4.

Freund, F. and Wengeler, H. (1982). The infrared spectrum of hydroxide-compensated defect sites in carbon-doped magnesium oxide and calcium oxide single crystals. *J. Phys. Chem. Solids*, **43**, 129–45.

Freund, F. (1983). The oxygen (1^-) state, hydrogen, and carbon in solid solution in refractory oxides. *High Temp.-High Pressures*, **15**, 335–46.

Freund, F., Wengeler, H. *et al.* (1983). Hydrogen and carbon derived from dissolved H_2O and CO_2 in minerals and melts. *Bull. Mineral.*, **106**, 185–200.

Freund, F. (1984). Hydrogen and nitrogen gas from magmatic rocks—a solid state viewpoint. *Oil Gas J.*, **82**, 140–1.

Freund, F. (1984). Volume instabilities in the mantle as a possible source for kimberlite formation. *Dev. Petrol.* **11A** (Kimberlites, Vol. 1) 405–15, 435–66.

Freund, F. (1986). Solute carbon and carbon segregation in magnesium oxide single crystals—a secondary mass spectrometry study. *Phys. Chem. Minerals.*, **13**, 262–76.

Freund, F. and Oberheuser, G. (1986). Water dissolved in olivine: a single crystal infrared study. *J. Geophys. Res. B*, **91**, 745–61.

Freund, F. (1986). Carbon in oxides and silicates—dissolution versus exsolution. *J. Crystal Growth*, **75**, 107–121.

Freund, F. (1986). Comment on Solute carbon and carbon segregation in magnesium oxide single crystals—a secondary ion mass spectrometry study by I. S. T. Tsong and U. Knipping. Reply. *Phys. Chem. Minerals.*, **13**, 280.

Freund, F. (1987). Hydrogen and carbon in solid solution in oxides and silicates.. *Phys. Chem. Minerals*, **15**, 1–18.

Frost, B. R. (1979). Mineral equilibriums involving mixed-volatiles in a carbon–oxygen–hydrogen fluid phase: the stabilities of graphite and siderite. *Am. J. Sci.*, **279**, 1033–59.

Galimov, E. M. (1973). Possibility of natural diamond synthesis under conditions of cavitation, occurring in a fast-moving magmatic melt. *Nature*, **243**, 389–91.

Garanin, V. K., Guseva, E. V., Dergachev, D. V., Kudryavtseva, G. P. and Orlov, R. Yu. (1988). Diamond crystals in garnets from granite–gneisses. *Dokl. Akad. Nauk SSSR*, **298**, 190–4.

Giardini, A. A. and Tydings, J. E. (1962). Diamond synthesis: observations on the mechanism of formation. *Am. Mineral.*, **47**, 1393–421.

Giardini, A. A., Salotti, C. A. and Lakner, J. F. (1968). Synthesis of graphite and hydrocarbons by reaction between calcite and hydrogen. *Science*, **159**, 317–19.

Giardini, A. A. and Melton, C. E. (1975). Gases released from natural and synthetic diamonds by crushing under high vacuum at 200°C and their significance to diamond genesis. *Fortschr. Min.*, **52**, 455–64.

Giardini, A. A., Hurst, V. J., Melton, C. E. and Stormer, J. C., Jr. (1974). Biotite as a primary inclusion in diamond: its nature and significance. *Am. Mineral.*, **59**, 783–9.

Glassley, W. E. (1983). Deep crustal carbonates as CO_2 fluid sources: evidence from metasomatic reaction zones. *Contrib. Mineral. Petrol.*, **84**, 15.

Goldsmith, J. R. (1986). The role of hydrogen in promoting aluminum–silicon interdiffusion in albite ($NaAlSi_3O_8$) at high pressures. *Earth Planet. Sci. Lett.*, **80**, 135–8.

Goldsmith, J. R. (1987). Aluminum/silicon interdiffusion in albite: effect of pressure and the role of hydrogen. *Contrib. Mineral. Petrol.*, **95**, 311–21.

Gorbachev, N. S., Zyryanov, V. N. and Boettcher, A. L. (1985). Solubility of sulfides in fluid-containing silicate melts at high pressures. *Ocherki Fiz.-Khim. Petrol.*, **13**, 153–65.

Haggerty, S. E. and Tompkins, L. A. (1983). Redox state of earth's upper mantle from kimberlite ilmenites. *Nature*, **303**, 295.

Haggerty, S. E. and Tompkins, L. A. (1984). Subsolidus reactions in kimberlitic ilmenites: exsolution, reduction and redox state of the mantle. *Dev. Petrol.*, **11A** (Kimberlites, Vol. 1), 335–57.

Haggerty, S. E. (1986). Diamond genesis in a multiply-constrained model. *Nature*, **320**, 34–8.

Harte, B. (1986). [Comment on the] Genesis of diamond: A mantle saga-distorted in the telling. *Am. Mineral.*, **71**, 1258; Reply by Meyer **71**, 1259–60.

Hervig, R. L., Smith, J. V., Steele, I. M., Gurney, J. J., Meyer, H. O. A. and Harris, J. W. (1980). Diamonds: minor elements in silicate inclusions: pressure-temperature implications. *J. Geophys. Res., B*, **85**, 6919–29.

Hirano, S. (1970). Effect of coexisting minerals on graphitization of carbon under pressure. Dr. Engr. thesis, Nagoya University.

Holloway, J. R. and Jakobsson, S. (1986). Volatile solubilities in magmas: transport of volatiles from mantles to planet surfaces. *J. Geophys. Res. B*, **91**, D505–8.

Homeny, J., Nelson, G. G. and Risbud, S. H. (1988). Oxycarbide glasses in the Mg–Al–Si–O–C system. *J. Am. Ceram. Soc.*, **71**, 386–90.

Itaya, T. (1981). Carbonaceous material in pelitic schists of the Sanbagawa metamorphic belt in central Shikoku, Japan. *Lithos*, **14**, 215–24.

Ivankin, P. F., Argunov, K. P. and Boris, E. I. (1983). Evolution of conditions for diamond formation in the kimberlite process. *Sov. Geol.*, 30–8.

Jacobs, G. K. and Kerrick, D. M. (1981). Methane: an equation of state with application to the ternary system H_2O–CO_2–CH_4. *Geochim. Cosmochim. Acta* **45**, 607–14.

Jakobsson, S. and Holloway, J. R. (1986). Crystal–liquid experiments in the presence of a carbon–oxygen–hydrogen fluid buffered by graphite + iron + wustite: experimental method and near-liquidus relations in basanite. *J. Volcanol. Geotherm. Res.*, **29**, 265–91.

Jedwab, J. and Boulegue, J. (1984). Graphite crystals in hydrothermal vents. *Nature*, **310**, 41–43.

Johnson, W. (1915). Origin and formation of the diamond. *S. African J. Sci.*, **II**, 275–86.

Kadik, A. A. (1988). Effect of melting on the evolution of fluid and oxidation–reduction regimes of the earth's upper mantle. *Geokhimiya*, **2**, 236–45.

Kamatsu, H. (1973). Letter from the depth of the earth. Diamonds, growth and nature. *Kotai Butsuri*, **8**, 293–304.

Kaminskii, F. V., Kulakova, I. I. and Ogloblina, A. I. (1985). Polycyclic aromatic hydrocarbons in carbonado and diamond. *Dokl. Akad. Nauk SSSR*, **283**, 985–8.

Karkhanis, S. N. (1977). Synthesis of abiogenic graphite under Precambrian conditions. *J. Geol. Soc. India*, **18**, 97–103.

Kathrein, H., Gonska, H. and Freund, F. (1983). Subsurface segregation and diffusion of carbon in magnesium oxide. *Appl. Physics A. Solids and Surfaces*, **30**, 33.

Kennedy, G. C. and Nordlie, B. E. (1968). The genesis of diamond deposits. *Econ. Geol.*, **63**, 495–503.

Kitamura, M., Kondoh, S., Morimoto, N., Miller, G. H., Rossman, G. R. and Putnis, A. (1987). Planar OH-bearing defects in mantle olivine. *Nature*, **328**, 143–5.

Kucha, H., Kwiecinska, B., Piestrzynski, A. and Wieczorek, A. (1979). On the genesis of graphite from magnetite rocks of Krzemianka (NE Poland). *Mineralogia Polonica*, **10**, 81–8.

Kulakova, I. I., Ogloblina, A. I., Rudenko, A. P., Florovskaya, V. N., Botkunov, A. I. and Skvortsova, V. L. (1982). Polycyclic aromatic hydrocarbons in diamond-associated minerals and possible mechanism of their formation. *Dokl. Akad. Nauk SSSR*, **267**, 1458–61.

Langford, R. E., Melton, C. E. and Giardini, A. A. (1974). Diamond growth by sulphide reduction of CO_2. *Nature*, **249**, 647.

Langford, R. E. (1978). The origin of diamonds II. Theoretical Study. *J. Korean Chem. Soc.* (*Taehan Hwakak Hoechi*), **22**, 138–49.

Lapin, A. V. and Marshintsev, V. K. (1984). Carbonatites and kimberlitic carbonatites. *Geol. Rudn. Mestorozhd.*, **26**, 28–42.

Larimer, J. W. and Bartholomay, M. (1979). The role of carbon and oxygen in cosmic gases: some applications to the chemistry and mineralogy of enstatite chondrites. *Geochim. Cosmochim. Acta*, **43**, 1455–66.

Letnikov, F. A. (1983). Formation of diamonds in deep-seated tectonic zones. *Dokl. Akad. Nauk SSSR*, **27**, 433–5.

Lukanin, O. A. and Kadik, A. A. (1987). Melting of ultrabasic mantle matter in the presence of oxidation–reduction conditions. *Vulkanol. Seismol.*, 3–13.

Luth, R. W. and Boettcher, A. L. (1986). Hydrogen and the melting of silicates. *Am. Mineral.*, **71**, 264–76.

Luth, R. W., Mysen, B. O. and Virgo, D. (1987). Raman spectroscopic study of the solubility behavior of hydrogen in the system sodium oxide–alumina–silica–hydrogen. *Am. Mineral.*, **72**, 481–6.

Mainwood, A. and Stoneham, A. M. (1984). Interstitial muons and hydrogen in diamond and silicon. *J. Phys. C.: Solid State Phys.*, **17**, 2513–24.

Malinovskii, I. Yu., Godovikov, A. A., Ran, E. A. and Logvinov, V. M. (1981). Study of silicate systems and development of superhigh pressure apparatus in connection with problems of mantle petrology and diamond genesis. *Eksperim. Petrol. Vysok. Davlenii, Novosibirsk*, 3–31.

Mal'kov, B. A. (1978). Conditions of diamond formation in nature according to crystallophysical data and the results of the experimental melting of peridotites. *Dokl. Akad. Nauk SSSR*, **243**, 469–72.

Mamchur, G. P., Mel'nik, Yu. M., Khar'kin, A. D. and Yarynych, O. A.

(1980). Origin of carbonates and bituminous matter in kimberlite pipes according to carbon isotope composition. *Geokhimiya*, 540–7.

Marakushev, A. A., Bezmen, N. I. and Mal'kov, B. A. (1981). Zoning of crystals in diamond-containing rock. *Mineral. Zh.*, **3**, 37–44.

Marakushev, A. A. (1981). Problem of the fluid regime in the formation of diamond-bearing rocks. *Geol. Rudn. Mestorozhd.*, **23**, 3–17.

Marakushev, A. A. (1985). Mineral associations of diamond and the problem of the formation of diamond-containing magmas. *Ocherki Fiz.-Khim. Petrol.*, 5–53.

Marx, P. C. (1972). Pyrrhotine and the origin of terrestrial diamonds. *Mineral Mag.*, **38**, 636–8.

Mathez, E. A., Dietrich, V. J. and Irving, A. J. (1984). The geochemistry of carbon in mantle peridotites. *Geochim. Cosmochim. Acta*, **48**, 1849–59.

Mathez, E. A., Blacic, J. D., Berry, J., Hollander, M. and Maggiore, C. (1987). Carbon in olivine: results from nuclear reaction analysis. *J. Geophys. Res. B*, **92**, 3500–6.

Mattioli, G. S. and Wood, B. J. (1986). Upper mantle fugacity recorded by spinel lherzolites. *Nature*, **322**, 626.

Melton, C. E. and Giardini, A. A. (1974). The composition and significance of gas released from natural diamonds from Africa and Brazil. *Am. Mineral.*, **59**, 775–82.

Melton, C. E. and Giardini, A. A. (1975). Experimental results and a theoretical interpretation of gaseous inclusions found in Arkansas natural diamonds. *Am. Mineral.*, **60**, 413–17.

Melton, C. E. and Giardini, A. A. (1981). The nature and significance of occluded fluids in three Indian diamonds. *Am. Mineral.*, **66**, 746–50.

Melton, C. E. and Giardini, A. A. (1982). The evolution of the earth's atmosphere and oceans. *Geophys. Res. Lett.*, **9**, 579–82.

Meyer, H. O. A. and Boyd, F. R. (1972). Composition and origin of crystalline inclusions in natural diamond. *Geochim. Cosmochim. Acta*, **36**, 1255–73.

Meyer, H. O. A. and Tsai, H. M. (1976). Mineral inclusions in diamond: temperature and pressure of equilibration. *Science*, **191**, 849–51.

Meyer, H. O. A. (1985). Genesis of diamond: a mantle saga. *Am. Mineral.*, **70**, 344–55.

Miller, G. H., Rossman, G. R. and Harlow, G. E. (1987). The natural occurrence of hydroxide in olivine. *Phys. Chem. Minerals*, **14**, 461–72.

Mitchell, R. H. and Crocket, J. H. (1971). Diamond genesis—a synthesis of opposing views. *Mineral. Deposita*, **6**, 392–403.

Miyashiro, A. (1964). Oxidation and reduction in the earth's crust with special reference to the role of graphite. *Geochim. Cosmochim. Acta*, **28**, 717–29.

Moore, A. E. (1987). A model for the origin of ilmenite in kimberlite and diamond: Implications for the genesis of the discrete nodule (megacryst) suite. *Contrib. Mineral. Petrol.*, **95**, 245–53.

Miyano, T. and Klein, C. (1986). Fluid behavior and phase relations in the system iron–magnesium–silicon–carbon–oxygen–hydrogen: application to high-grade metamorphism of iron formations. *Am. J. Sci.*, **286**, 540–75.

Mysen, B. O., Eggler, D. H., Seitz, M. G. and Holloway, J. R. (1976).

Carbon dioxide in silicate melts and crystals. Part I. Solubility measurements. *Am. J. Sci.*, **276**, 455–79.

Mysen, B. O. (1977). The solubility of H_2O and CO_2 under predicted magma genesis conditions and some petrological and geophysical implications. *Rev. Geophys. Space Phys.*, **15**, 351–61.

Mysen, B. O. and Virgo, D. (1980). Solubility mechanisms of carbon dioxide in silicate melts: a Raman spectroscopic study. *Am. Mineral.*, **65**, 885–99.

Mysen, B. O., Virgo, D., Harrison, W. J. and Scarfe, C. M. (1980). Solubility mechanisms of H_2O in silicate melts at high pressures and temperatures: a Raman spectroscopic study. *Am. Mineral.*, **65**, 900–14.

Mysen, B. O. and Virgo, D. (1982). Solubility mechanisms of water in silicate melts at high pressures and temperatures: Raman spectroscopic study: reply. *Am. Mineral.*, **67**, 155.

Nestor, L. R. (1967). Glass containing dissolved carbon, methods of making and using and obtaining graphite. U.S. Patent 3,348,917: filed 7/22/60; issued 10/24/67.

Neuhaus, A. (1954). Uber die Synthese des Diamanten. *Angew. Chem.*, **66**, 525–36.

Nickel, K. G. and Green, D. H. (1985). Empirical geothermobarometry for garnet peridotites and implications for the nature of the lithosphere, kimberlites and diamonds. *Earth Planet. Sci. Lett.*, **73**, 158–70.

Nikolayeva, O. V. and Germanov, A. I. (1972). Thermodynamic equilibria in the system $C-H_2O$ under hydrothermal conditions. *Dokl. Akad. Nauk SSSR*, **207**, 958–61.

Nikol'skii, N. S. (1981). Metastable crystallization of natural diamonds from the fluid phase. *Dokl. Akad. Nauk SSSR*, **256**, 954–8.

Nikol'skii, N. S. (1982). Modelling equilibrium compositions of multicomponent fluid phases (as in the system H–O–C) and their importance in magmatic activity. *Dokl. Akad. Nauk SSSR*, **257**, 134–8.

Nikol'skii, N. S. (1984). Crystallization conditions for some reduced mineral phases and their petrogenetic importance. *Vulkanol. Seismol.*, 45–58.

Nuth, J. A. (1987). Are small diamonds thermodynamically stable in the interstellar medium? *Astrophys. Space Sci.*, **139**, 103–9.

Ohmoto, H. and Kerrick, D. (1977). Devolatilization equilibria in graphitic systems. *Am. J. Sci.*, **277**, 1013–44.

Olsen, E. and Fuchs, L. (1968). Krinovite, $NaMg_2CrSi_3O_{10}$: a new meteorite mineral. *Science*, **161**, 786–7.

Ostrovsky, I. A. (1979). The thermodynamics of substances at very high pressures and temperatures and some mineral reactions in the earth's mantle. *Phys. Chem. Minerals*, **5**, 105–18.

Pasteris, J. D. (1981). Occurrence of graphite in serpentinized olivines in kimberlite. *Geology*, **9**, 356–9.

Pasteris, J. D. (1984). Kimberlites: complex mantle melts. *Ann. Rev. Earth Planet. Sci.*, **12**, 133–53.

Patel, A. R. and Kuruvilla, A. (1984). On the possible origins of natural diamonds. *Pramana*, 22377–86.

Perchuk, L. L. and Suvorova, V. A. (1973). Thermodynamic calculation of the fugacities of carbon monoxide and carbon dioxide in the graphite-diamond phase transition region. *Ocherki Fiz.-Khim. Petrologii,* No. 3, 15–18.

Perchuk, L. L. and Vagonov, V. I. (1980). Petrochemical and thermodynamic evidence on the origin of kimberlites. *Contrib. Mineral. Petrol.,* **72,** 219–28.

Petrov, V. S. (1967). On the natural genesis of diamonds. *Rost Kristallov,* **7,** Part 1, 105–11.

Petrov, V. S. (1972). Equilibrium in the olivine–diamond system. *Rost Kristallov,* **9,** 73–5.

Pokhilenko, N. P., Sobolev, N. V., Sobolev, V. S. and Lavrent'ev, Yu. G. (1976). Xenolith of diamond-containing ilmenite–pyrope lherzolite from the 'Udachnaya' kimberlite pipe (Yakutia). *Dokl. Akad. Nauk SSSR,* **231,** 438–41.

Popivnyak, I. V., Demin, B. G., Levitskii, V. V. and Koptil, V. I. (1980). New data on volatile components of mantle mineral-forming media. *Dokl. Akad. Nauk SSSR,* **254,** 1238–41.

Portnov, A. M. (1982). Self-oxidation of mantle fluid and the genesis of diamond of kimberlite. *Dokl. Akad. Nauk. SSSR,* **267,** 942.

Rai, C. S., Sharma, S. K., Muenow, D. W., Matson, D. W. and Byers, C. D. (1983). Temperature dependence of carbon dioxide solubility in high pressure quenched glasses of diopside composition. *Geochim. Cos-mochim. Acta,* **47,** 953–8.

Richardson, S. H., Gurney, J. J., Erlank, A. J. and Harris, J. W. (1984). Origin of diamonds in old enriched mantle. *Nature,* **310,** 198–202.

Rimbach, H. and Chatterjee, N. D. (1987). Equations of state for hydrogen, water, and hydrogen–water fluid mixtures at temperatures above 0.01 degree C and at high pressures. *Phys. Chem. Minerals,* **14,** 560–9.

Ringwood, A. E. (1977). Synthesis of pyrope–knorringite solid solution series. *Earth Planet. Sci. Lett.,* **36,** 443–8.

Robinson, D. N. (1978). The characteristics of natural diamond and their interpretation. *Minerals Sci. Engrg,* **10,** 55–72.

Rodewald, H. J. (1960). *The genesis of diamonds.* Verlag Meier and Cie, Schaffhausen, Switzerland.

Rosenhauer, M., Woermann, E., Knecht, B. and Ulmer, C. G. (1977). The stability of graphite and diamond as a function of the oxygen fugacity of the mantle, Extended Abstract. Second International Kimberlite Conference.

Rumble, D., Duke, E. F. and Hoering, T. L. (1986). Hydrothermal graphite in New Hampshire: evidence of carbon mobility during regional metamorphism. *Geology,* **14,** 452–5.

Ryabchikov, I. D. (1980). Nature of kimberlite 'magmas'. *Geol. Rudn. Mestorozhd.,* **22,** 18–26.

Ryabchikov, I. D. (1982). Hydrothermal reactions in the earth's mantle. In *Proceedings of the 1st International Symposium on Hydrothermal Reactions.* Tokyo, pp. 244–57.

Ryabchikov, I. D., Ukhanov, A. V. and Ishii, T. (1985). Redox equilibria in the ultramafic basic rocks from the upper mantle of the Yakutian kimberlite province. *Geokhimiya,* **8,** 1110–23.

Samoilovich, M. I. (1987). Thermodynamics of diamond formation. *Dokl. Akad. Nauk SSSR*, **296**, 602–4.

Schrocke, H. (1969). New aspect of diamond formation. *Z. Dt. Gemmol. Ges.*, 1851–5.

Schulze, D. J. (1986). Calcium anomalies in the mantle and a subducted metaserpentinite origin for diamonds. *Nature*, **319**, 483–5.

Sellschop, J. P. F. (1975). Evidence on the environment of diamond genesis from trace element studies of natural diamonds. *Diamond Research*, 35–41.

Serebryanaya, N. R., Losev, V. G., Voronov, O. A., Rakhmanina, A. V. and Yakovlev, E. N. (1985). Morphology of diamond crystals synthesized from hydrocarbons. *Kristallografiya*, 1026–7.

Sharp, W. E. (1966). Pyrrhotite: a common inclusion in South African diamonds. *Nature*, **211**, 402–3.

Simakov, S. K. (1982). Formation and crystallization of diamond in a mantle melt from fluid. *Dokl. Akad. Nauk SSSR*, **266**, 470–3.

Simakov, S. K. (1983). Formation of carbon in mantle fluid from the reaction of nitrogen with methane. *Dokl. Akad. Nauk SSSR*, **268**, 206–10.

Simakov, S. K. (1984). Possibility of diamond metastable formation from fluids under conditions in the earth's crust. *Dokl. Akad. Nauk SSSR*, **278**, 953–7.

Simakov, S. K. (1987). Diamond formation in the process of kimberlite magma evolution. *Dokl. Akad. Nauk SSSR*, **293**, 681–4.

Slawson, C. B. (1953). Synthesis of graphite at room temperature. *Am. Mineral.*, **38**, 50–5.

Slodkevich, V. V. (1981). Some genetic aspects of diamond formation in a magmatic system of closed type. *Samorodn. Mineraloobraz. v Magmatich. Protesse. Materialy Konf. Yakutsk 1981*, pp. 149–53.

Slodkevich, V. V. (1987). Exadiamondiferous phlogopite lherzolites. *Dokl. Akad. Nauk SSSR*, **297**, 942–5.

Smyth, J. R. (1987). Beta-Mg_2SiO_4: A potential host for water in the mantle? *Am. Mineral.*, **72**, 1051–5.

Sobolev, N. V., Pokhilenko, N. P. and Efimova, E. S. (1984). Xenoliths of diamond-bearing peridotites in kimberlites and the problem of diamond origin. *Geol. Geofiz.*, 63–80.

Sobolev, N. V. (1983). Diamond paragenesis and the problem of abyssal mineral formation. *Zap. Vses. Mineral. O-va.*, **112**, 389–97.

Sobolev, V. V. (1987). Diamond crystallization in nature. *Fiz. Goreniya Vzryva*, **23**, 91–5.

Sobolev, N. V. and Shatskii, V. S. (1987). Carbon mineral inclusions in garnets of metamorphic rocks. *Geol. Geofiz.*, **7**, 77–80.

Specius, Z. and Bulanova, G. P. (1987). Native iron in diamondiferous eclogites from the Udachnaya kimberlite pipe. *Dokl. Akad. Nauk SSSR*, **294**, 1445–8.

Sunagawa, I. (1969). *Diamonds—Their Genesis and Properties*. Maruzen, Tokyo.

Taylor, W. R. and Green, D. H. (1988). Measurement of reduced peridotite-C–O–H solidus and implications for redox melting of the mantle. *Nature*, **332**, 349–52.

Trofimov, V. S. (1972). Origin of diamonds. *Geol. Zh.*, **32**, 146–8.

Tsong, I. S. T., Knipping, U., Loxton, C. M. and Magee, C. W. (1985). Carbon on surfaces of magnesium oxide and olivine single crystals. Diffusion from the bulk or surface contamination? *Phys. Chem. Minerals*, **12**, 261–70.

Tsong, I. S. T. and Knipping, U. (1986). 'Solute carbon and carbon segregation in magnesium oxide single crystals—a secondary ion mass spectrometry study' by F. Freund. Comments. *Phys. Chem. Minerals*, **13**, 277–9.

Tuttle, D. L. (1973). Inclusions in Herkimer 'diamond'. *Lapidary J.*, Sept., 966–76.

Voznyak, D. K., Gritsyk, V. V., Kvasnitsa, V. N. and Galaburda, Yu. A. (1973). Inclusions of petroleum in Marmarosh diamonds. *Dopov. Akad. Nauk Ukr. RDR, Ser. B.*, **35**, 1059–62.

Watson, E. B., Sneeringer, M. A. and Ross, A. (1982). Diffusion of dissolved carbonate in magmas; experimental results and applications. *Earth and Planet. Sci. Lett.*, **61**, 346.

Watson, E. B. (1986). Immobility of reduced carbon along grain boundaries in dunite. *Geophys. Res. Lett.*, **13**, 529–32.

Weis, P. L., Friedman, I. and Gleason, J. P. (1981). The origin of epigenetic graphite: evidence from isotopes. *Geochim. Cosmochim. Acta*, **45**, 2325–32.

Wenlandt, R. F. and Mysen, B. O., Melting phase relations of natural peridotite + CO_2 as a function of degree of partial melting at 15 and 30 kb. *Am. Mineral*, **65**, 37–44 (1980).

Wentorf, R. H., Jr. and Bovenkerk, H. P. (1961). On the origin of natural diamonds. *Astrophys. J.*, **134**, 995–1005.

Wentorf, R. H., Jr. (1966). Solutions of carbon at high pressure. *Ber. Bun. fur Phys. Chem.*, **70**, 975–82.

Wintsch, R. P., O'Connell, A. F., Ransom, B. L. and Wiechmann, M. J. (1981). Evidence for the influence of $f(CH_4)$ on the crystallinity of disseminated carbon in greenschist facies rocks, Rhode Island, U.S.A. *Contrib. Mineral. Petrol.*, **77**, 207–13.

Woermann, E., Knecht, B. and Rosenhauer, M. (1981). Diamond synthesis, U.S. Patent 4,254,091; filed 8/16/79; issued 3/3/81.

Woermann, E., Knecht, B., Rosenhauer, M. and Ulmer, G. C. (1977). The stability of graphite in the system C–O. Extended Abstract. *Second International Kimberlite Conference*, 1977.

Woermann, E., Knecht, B., Rosenhauer, M. and Ulmer, G. C. (1978). Kohlenstoff-Karbonatgleichgewichte und der Redoxzustand im Erdmantel. *Fortschr. Mineralogie Beiheft*, **56**, 144–5.

Woods, G. S. and Collins, A. T. (1983). Infrared absorption spectra of hydrogen complexes in type I diamonds. *J. Phys. Chem. Solids*, **44**, 471–5.

Wyllie, P. J. and Tuttle, O. F. (1959). Synthetic carbonatite magma. *Nature*, **183**, 770.

Wyllie, P. J. and Tuttle, O. F. (1959). Melting of calcite in the presence of water. *Am. Mineral.*, **44**, 453–9.

Wyllie, P. J. (1979). Magmas and volatile components. *Am. Mineral.*, **64**, 469–500.

Wyllie, P. J., Huang, W. L., Otto, J. and Byrnes, A. P (1984). Carbonation of peridotites and decarbonation of siliceous dolomites represented in the system calcium oxide–magnesium oxide–silicon dioxide–carbon dioxide to 30 kbar. *Tectonophysics*, **100**, 359–88.

Wyllie, P. J. (1987). Discussion of recent papers on carbonated peridotite, bearing on mantle metasomatism and magmatism. *Earth Planet. Sci. Lett.*, **82**, 391–7.

Ziegenbein, D. and Johannes, W. (1980). Graphite in C–H–O fluids: an unsuitable compound to buffer fluid composition at temperatures up to 700°C. *Neues Jahrb. Min., Mh.*, **8**, 289–305.

Ziegenbein, D. and Johannes, W. (1982). Activities of carbon dioxide in supercritical carbon dioxide–water mixtures, derived from high-pressure mineral equilibrium data, *High-Pressure Res. Geosci., Results Priority Program Proc. Its Final Colloq.* ed. W. Schreyer, Schweizerbart, Stuttgart, pp. 493–500.

Zimin, S. S. and Zalishchak, B. L. (1986). New model of the formation of carbonatites and ores related to them. *Dokl. Akad. Nauk SSSR*, **289**, 700–2.

11

Hydrothermal Preparation of Fine Powders

SHIGEYUKI SŌMIYA

The Nishi Tokyo University, Tokyo, Japan

ABSTRACT

There are several methods for producing very fine powders under hydrothermal conditions, including: (1) hydrothermal oxidation, (2) hydrothermal precipitation, (3) hydrothermal crystallization, (4) hydrothermal synthesis, (5) hydrothermal decomposition, (6) hydrothermal dehydration, (7) hydrothermal anodic oxidation, (8) reactive electrode submerged arc, (9) hydrothermal mechanochemical reaction. This chapter describes these processes.

1. INTRODUCTION

Fine powder preparation is becoming an important process in recent high-technology ceramics. Preparation methods for fine powders can be divided into two categories:

- breakdown (size reduction) processes;
- buildup (size increase) processes.

In the breakdown processes large grains are broken physically and mechanically into small grains by crushing, grinding, or milling. The buildup methods have been developed recently for fine powders. The fine particles are made from smaller particles chemically, e.g. by precipitation, sol–gel process, hydrothermal preparation, and so on.

For ideal powders, there are several requirements, including.

(1) fine particle less than 1 μm;
(2) narrow particle size distribution;
(3) little or no macroscopic agglomeration;
(4) homogeneity;
(5) controllable composition and purity;
(6) controlled microstructure;
(7) controlled mechanical properties.

One of the possible approaches is hydrothermal processing. There are several methods for preparing fine powders under hydrothermal conditions; for example, under conditions of high temperature and high pressure in aqueous solution over 100°C and 1 bar.

(1) hydrothermal oxidation
(2) hydrothermal precipitation
(3) hydrothermal crystallization
(4) hydrothermal synthesis
(5) hydrothermal decomposition
(6) hydrothermal dehydration
(7) hydrothermal anodic oxidation
(8) reactive electrode submerged arc (RESA)
(9) hydrothermal mechanochemical reaction

Hydrothermal hydrolysis is similar to hydrothermal precipitation.

According to Roy (private communication, 1988), the term 'hydrothermal synthesis' involves H_2O as a catalyst (and occasionally as a component of the solid phase) in synthesis at elevated temperature (>100°C) and pressure (>a few atmospheres).

Many good summary papers have appeared in the literature. Examples include Morey [1], Roy and Tuttle [2], Laudise and Nielsen [3], Laudise [4, 5], Rabenau [6, 7], Ballman and Laudise [8], Demianets *et al.* [9]. This chapter does not aim to describe all the review papers and scientific papers that have appeared, and consequently numerous useful reviews and papers are not mentioned.

2. EQUIPMENT FOR HYDROTHERMAL SYNTHESIS

A high-temperature high-pressure vessel (an autoclave) is necessary to maintain the required temperatures and pressures. According to

TABLE 1
Limits of pressure and temperature of autoclave[a]

	Pressure (Mpa)	Temperature (°C)
Pyrex (5 mm inside diameter, 9 mm outside diameter)	0·6	250
Quartz (5 mm inside diameter, 9 mm outside diameter)	0·6	300
Morey type	40	400
Welded Walker–Buehler	200	480
Delta Ring	230	400
Bridgman seal	370	500
Modified Bridgman seal	370	500
Cold-seal test-tube	400	200
Stellite 25	200	800
Rene 41	100	740
TZ M	300	1100

[a] Modified from Ref. 3.

Laudise and Nielson [3] and S. Sōmiya [10, 24], the limits of temperature and pressure are as shown in Table 1.

Recently developed sealing methods are as follows:

- Flat plate closure (Morey): Fig. 1
- Cold-cone seat closure (Tuttle): Fig. 2
- Welded closure (Walker and Buehler): Fig. 3
- Full Bridgman closure (Bridgman): Fig. 4
- Delta and 'D' ring closure: Fig. 5
- Modified Bridgman closure: Fig. 6
- Grayloc closure: Fig. 7

Detailed explanation has appeared in the papers by Laudise and Nielsen [3] and by Asahara *et al.* [11].

Autoclave materials are important from the standpoint of experimental conditions such as corrosion, temperature, pressure and time limitation for hydrothermal reactions. Alloys for autoclaves are shown in Table 2. The life of the autoclave is dependent on alloy composition and properties, temperature, pressure, autoclave design, and duration of runs. An example of autoclave temperature–pressure–experiment duration is shown in Fig. 8.

The Morey bomb (Fig. 1) and Tuttle type (Fig. 2) are the most commonly used autoclaves for hydrothermal synthesis in many labora-

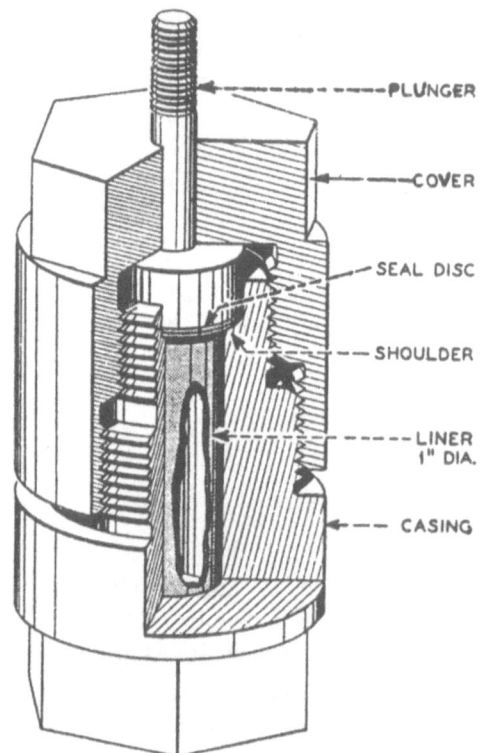

FIG. 1. Autoclave with flat plate closure from Ref. 3.

tories worldwide. The early versions of the Morey bomb were sealed
off with a fixed volume of water and the pressure was therefore fixed
by the temperature and water volume. This is one limitation of this
bomb. Secondly, casing and cover are fixed at a high temperature for a
long duration in a furnace. Consequently, in some cases one cannot
open the bomb after the experiment. Thirdly, temperature and
pressure limitations are 400°C and 50 MPa respectively, using environmentally resistant steel. Although this bomb has these disadvantages,
most laboratories have used it for preparing large amounts of samples.

The Tuttle type vessel (Fig. 2) is also called a test-tube or a cold-seal
pressure vessel. Pressure is maintained by a cone-in-cone arrangement. This vessel is simply a cylinder of metal with an axial hole
drilled from one end to within about 2·5 cm of the other end. The

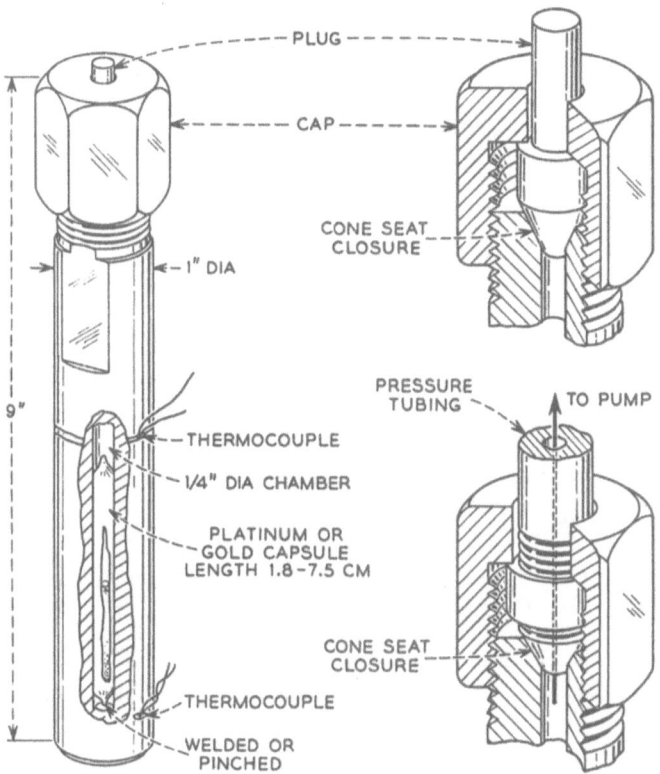

FIG. 2. Reaction vessel with a cold-cone seat closure, Tempress Inc., State College, Pennsylvania, USA (from Ref. 8).

closed end is placed upwards in a furnace with a pressure pump connection located on the outside of the furnace. The author's laboratory uses Pt wire or Ni–Cr wire for the contact between the thermocouple and the test-tube. Owing to changing design of the autoclave, after treatment it is very difficult to disassemble the test tube. Temperatures are measured with a thermocouple and pressure is detected by a pressure gage.

We use stainless steel stellite 25, and incomel X heat-resistant steels, and Ni–Cr–Mo alloys, as are shown in Table 2. The safety limits of these alloys differ from alloy to alloy and depend on temperature, pressure and duration. As already mentioned, it is safer and better with the thermocouple well outside the bomb.

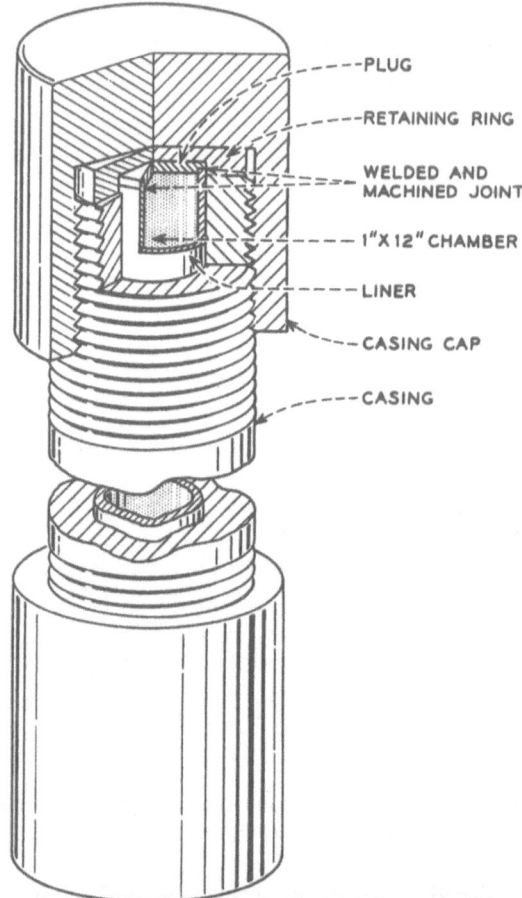

FIG. 3. Welded closure vessel (from Ref. 8).

After a new TZM alloy was developed a new design of autoclave appeared. This is one of the test-tube types. We are able to use this for high-temperature high-pressure experiments and it is shown in Fig. 9. Modifications of the test-tube type autoclave are shown in Figs 10 and 11. One of these is of floating type. Inserts are made of carbon-free iron, copper, silver (for alkaline solutions), titanium, platinum, glass or fused-quartz (for acid solutions); Teflon is used for all solutions, but

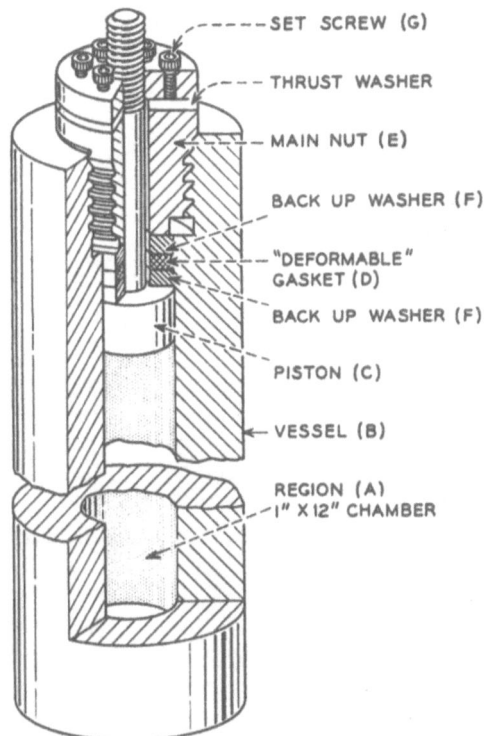

SET SCREW (G)

THRUST WASHER

MAIN NUT (E)

BACK UP WASHER (F)

"DEFORMABLE" GASKET (D)

BACK UP WASHER (F)

PISTON (C)

VESSEL (B)

REGION (A)
1" X 12" CHAMBER

FIG. 4. Full Bridgman closure (from Ref. 8).

at a temperature as low as 200°C. Figure 11 shows the so-called 'two-zone process'. The starting materials are placed in different solvents in separated zones, because the solubilities of the starting materials are different from solution to solution. This method has been developed for the preparation of Sb_2O_4, and for compounds of the type $ABO_4(A = Sb(III), Bi(III), B = Ta(V), Sb(V), Nb(V))$.

3. PROCESSING OF POWDER PREPARATION UNDER HYDROTHERMAL CONDITIONS

As mentioned in Section 1, there are several hydrothermal processes.

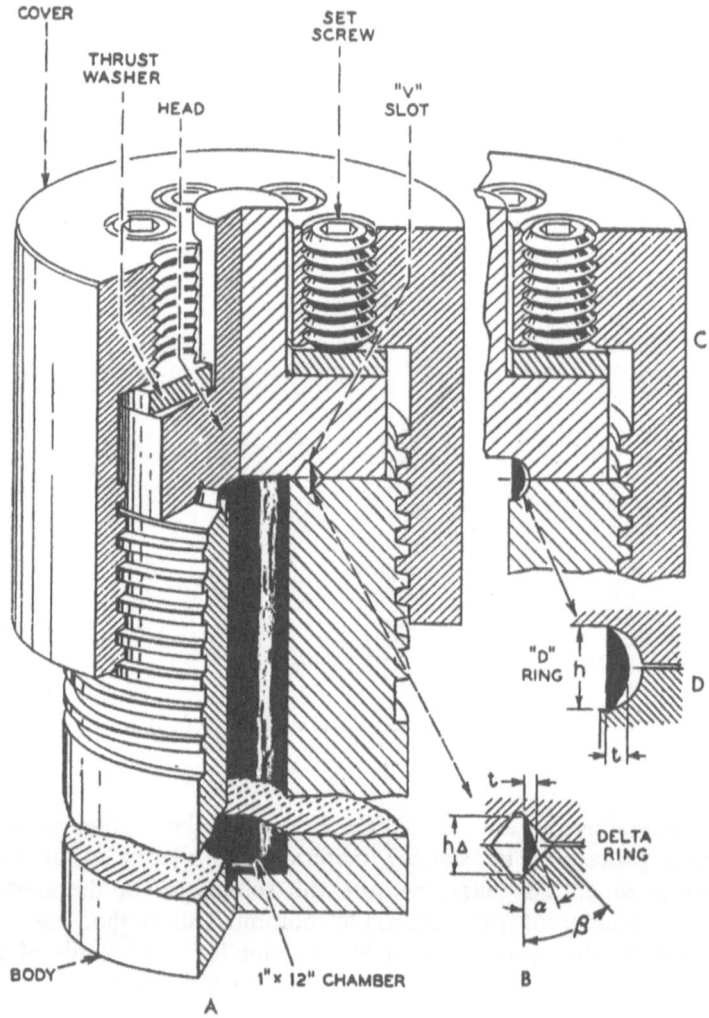

Fig. 5. Delta and 'D' ring vessels (American Instrument Company, Silver Spring, Maryland, USA, and G.E. Ltd, Wembley, England) (from Ref. 3).

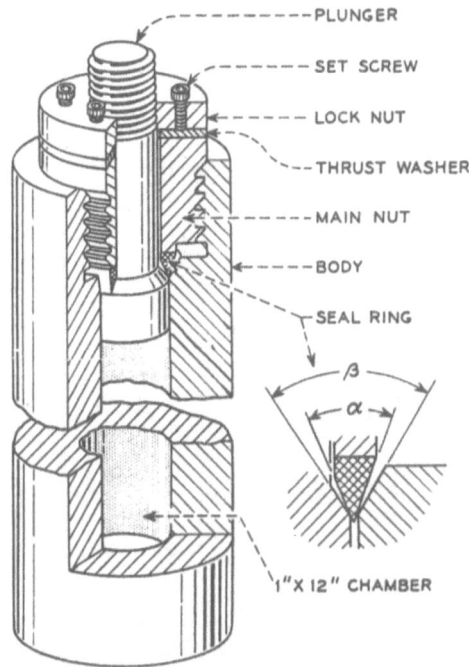

FIG. 6. Modified Bridgman vessel (Autoclave Engineers, Pennsylvania, USA) (from Ref. 8).

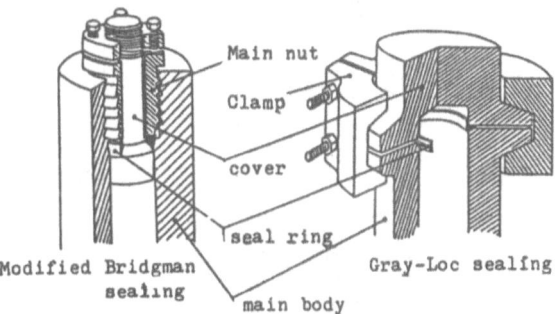

FIG. 7. Comparison for self- and pressure-energized sealing systems applied to larger autoclaves (modified Bridgman seal and Gray-Loc seal) (from Ref. 11).

Shigeyuki Sōmiya

TABLE 2
Alloys for autoclaves[a]

	C	Mn	Si	Cr	Ni	Co	Mo	W	Nb	Ti	Al	Others
Low-carbon steel	0·15	0·5										
4140 Steel	0·4	0·9	0·2	1			0·2					
Stainless Steel 304	0·08	0·6	0·6	15	9							
19-9-DL	0·3	1·1	0·6	18	9		1·2	1·2	0·4	0·3		
Croloy 15-5-N	0·10	1·5	0·5	15	15		1·5	1·5	1·0			N 0·12
Timken 17-22-A	0·30	0·59	0·66	1·25	0·25		0·51					Cu 0·14
Inconel X	0·05	0·5	0·4	14·5	Bal	1·0				2·5	0·22	Fe?
Udimet 500	0·08	0·2	0·2	20	Bal	15	3·5			3	0·8	
Stellite 25				19–21	9–11			14–16			3	
Rene 25				19	Bal	11	10					
TZM							Bal			0·5		Zr 0·08

[a] Modified from Ref. 3.

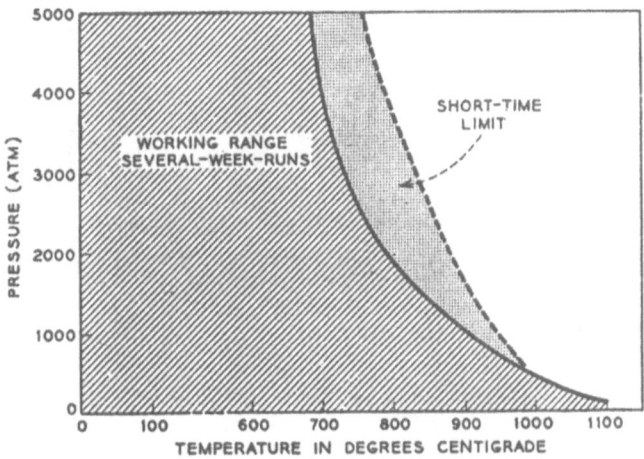

FIG. 8. *P–T* limits for satellite vessel (Catalog Information, Tempress Inc., State College, Pennsylvania, USA). (from Ref. 3).

3.1. Hydrothermal Oxidation

There are four types of hydrothermal oxidation reactions, namely,

(1) direct oxidation
(2) through hydroxide
(3) through hydride
(4) mixed type

Examples of reactions are as follows:

$Zr + H_2O \rightarrow ZrO_2 + H_2$	Type 3
$Fe + H_2O \rightarrow Fe_3O_4 + H_2$	Type 1
$Cr + H_2 \rightarrow Cr_2O_3 + H_2$	Type 1
$Ti + H_2O \rightarrow TiO_2 + H_2$	Type 3
$Nb + H_2O \rightarrow Nb_2O_5 + H_2$	Type 3
$Zn + H_2O \rightarrow ZnO + H_2$	Type 1
$Al + H_2O \rightarrow Al_2O_3 + H_2$	Type 2
$Hf + H_2O \rightarrow HfO_2 + H_2$	Type 3
$(Al + Zr) + H_2O \rightarrow (Al_2O_3 + ZrO_2) + H_2$	Type 4
$(Al + Hf) + H_2O \rightarrow (Al_2O_3 + HfO_2) + H_2$	Type 4

3.1.1. Oxidation of Ti Metals [18]

3.1.1.1. Experimental procedures. Titanium metal powder is used as starting material. Specimens for experiments were prepared in (a) a

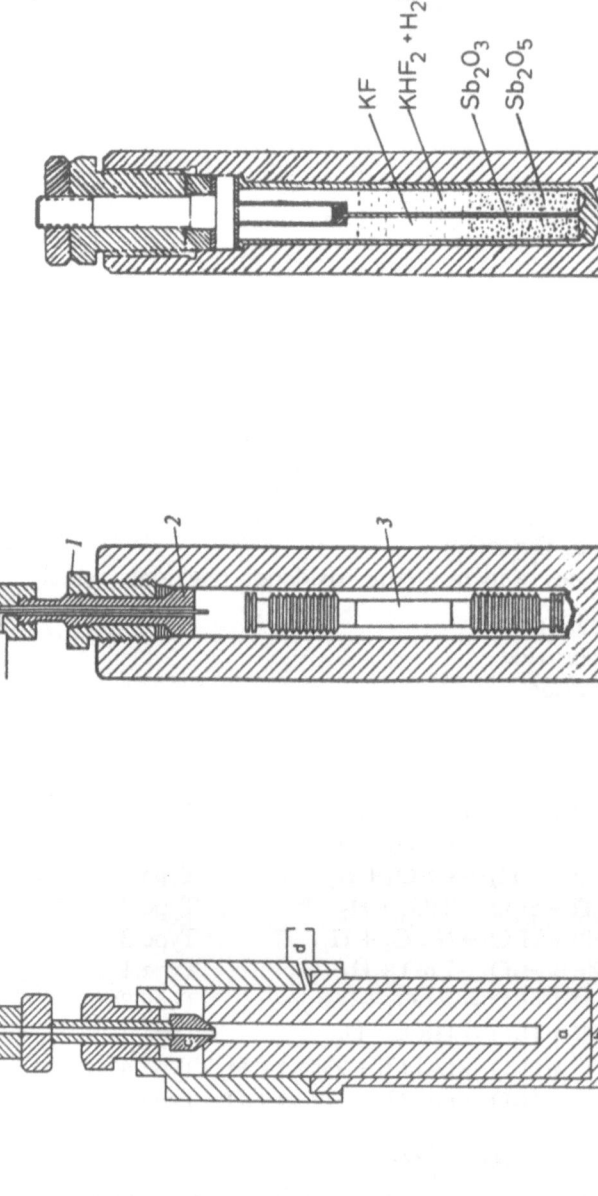

Fig. 9. TZM autoclave with inert-gas flushing. (a) Autoclave body of TZM (molybdenum) alloy. (b) Protective mantle of Inconel 750 (nickel alloy). (c) Conical seal. (d) Inlet vent for inert gas (from Ref. 7).

Fig. 10. Autoclave with a floating-type, variable-volume insert: 1, lock nut; 2, plunger; 3, insert (from Ref. 9).

Fig. 11. Two-section liner for separation of starting materials in different solvents (from Ref. 7).

closed system, that is, metal powder and redistilled water were sealed in a gold capsule (I.D. of 2·7 mm and length of 3·5 mm) in the molar ratio of $H_2O/Ti = 2/1$, or (b) an open system where the powder was set at the bottom of a gold capsule (I.D. of 2·7 or 4·6 mm and length of 20–40 mm) without sealing the top. Closed system implies that both ends of the capsule were welded and open system that only one end of the capsule was welded. In the open system, therefore, metals could react with excess water. Then, the capsule was heated in a test-tube type pressure vessel under the temperature, pressure and time conditions of 200–700°C, 10–150 MPa and 0–120 h, respectively.

After treatment, the pressure vessel was quenched in water and solid products in the capsule were examined by X-ray powder diffraction and scanning electron microscopy.

3.1.1.2. Results. Figures 12 and 13 illustrate variations of products with temperature under 100 MPa for 3 h in the closed and the open systems, respectively. As shown in these figures, Ti changed into titanium hydride (TiH_x, $x = 1·924$), anatase and rutile above 450°C. Over 600°C, all of Ti and anatase changed into rutile and TiH_x, and rutile was the only product at 700°C. Variations of products with time at 500°C under 100 MPa in the closed system and the open system are shown in Figs 14 and 15, respectively. Under these conditions, Ti changed into TiH_x, anatase and rutile within 8 h. After that, reactions from TiH_x and anatase to rutile continued. Little effect of pressure was observed in either system (Fig. 13).

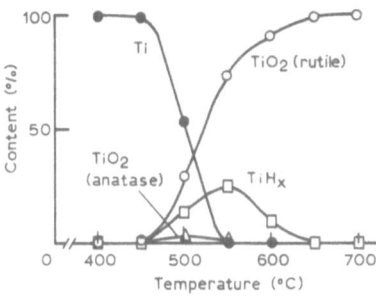 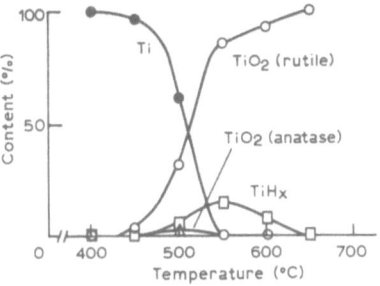

FIG. 12. Variation of amount of products with temperature by the hydrothermal oxidation of Ti in the closed system under 100 MPa for 3 h (from Ref. 18).

FIG. 13. Variation of amount of products with temperature by the hydrothermal oxidation of Ti in the open system under 100 MPa for 3 h (from Ref. 18).

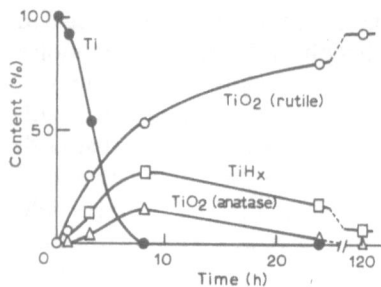

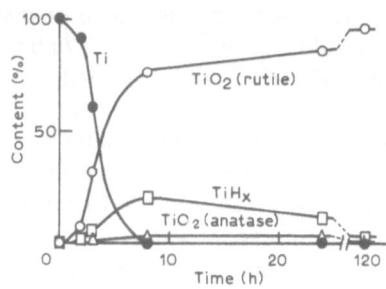

Fig. 14. Variation of amount of products with time by the hydrothermal oxidation of Ti in the closed system at 500°C under 100 MPa (from Ref. 18).

Fig. 15. Variation of amount of products with time by the hydrothermal oxidation of Ti in the open system at 500°C under 100 MPa (from Ref. 18).

From these results, the reactions of hydrothermal oxidation of Ti can be described as follows.

$$Ti + 2H_2O \rightarrow TiO_2 + 2H_2 \tag{1}$$

$$Ti + \frac{x}{2}H_2 \rightarrow TiH_x \quad (x = 1.924) \tag{2}$$

$$TiH_x + 2H_2O \rightarrow TiO_2 + \left(2 + \frac{x}{2}\right)H_2 \tag{3}$$

Initially, the surface of Ti is subjected to oxidation (reaction (1)); then TiH_x is formed by reaction with H_2 produced in reaction (1). This TiH_x changes into TiO_2 by reaction with water (reaction (3)).

From Figs 12–16, it is apparent that there exist differences between the closed system and the open system. One difference is the amount

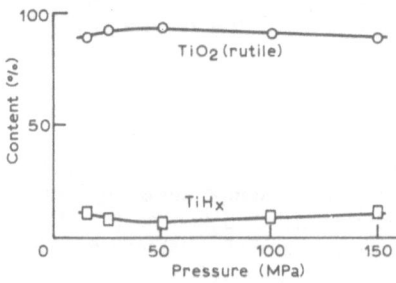

Fig. 16. Variation of amount of products with pressure by the hydrothermal oxidation of Ti in the closed system at 600°C for 3 h (from Ref. 18).

of TiH_x: in the closed system, more TiH_x was formed than in the open system. This can be attributed to higher hydrogen partial pressure (p_{H_2}) in the closed system than in the open system. In the closed system, hydrogen produced by oxidation of Ti cannot leave the capsule, while hydrogen does leave the capsule and p_{H_2} becomes much lower in the open system. Also in reaction (3), high p_{H_2} prevents oxidation of TiH_x; as a result, more TiH_x is produced in the closed system than in the open system.

3.2. Hydrothermal Precipitation [19, 20, 21]

Preparation of ZrO_2 and Y_2O_3 doped ZrO_2 powder is given as an example.

3.2.1. Experimental

$ZrOCl_2 \cdot 8H_2O$, $YCl_3 \cdot 6H_2O$ and $CO(NH_2)_2$ were used as starting materials. Experimental procedures are shown in Fig. 17.

3.2.2. Results

Well-crystallized 3Y-PSZ powder of 11·6 nm crystallite size was obtained by hydrothermal treatment at 220°C under 7 MPa for 5 h. The crystallite size was determined by using the Scherrer–Warrens equation on the half-width of the (111) line of tetragonal/cubic ZrO_2. The crystallite size decreased from 15·0 to 11·6 nm on increasing the temperature from 160° to 220°C, as shown in Fig. 18. However, the surface area (BET) of the products remained constant at 100 m²/g regardless of the temperature.

These powders appear to have consisted of metastable-cubic zirconia and a small amount of monoclinic zirconia. The content of the monoclinic phase decreased with increasing temperature under hydrothermal conditions. The metastable-cubic phase was determined by separating the (400) line of the X-ray diffraction patterns. The metastable-cubic phase was transformed into the tetragonal phase by calcination at temperatures above 800°C, as shown in Fig. 19.

3.3. Hydrothermal Crystallization [19, 20, 21]

3.3.1. Experimental procedure

The starting material was amorphous hydrous zirconia prepared from $ZrCl_4$ solution with 3 N NH_4OH, filtered, dispersed in distilled water

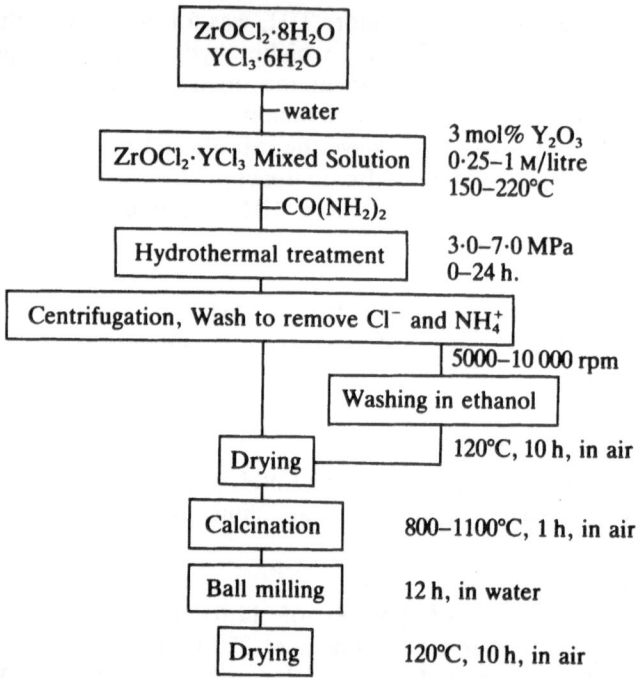

FIG. 17. Processing flow sheet of the hydrothermal homogeneous precipitation method (from Ref. 19).

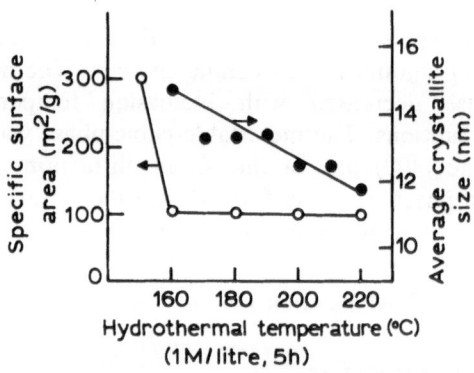

FIG. 18. Variation of surface area and crystallite size of PSZ (3 mol% Y_2O_3) powders with temperature under hydrothermal conditions (from Ref. 20).

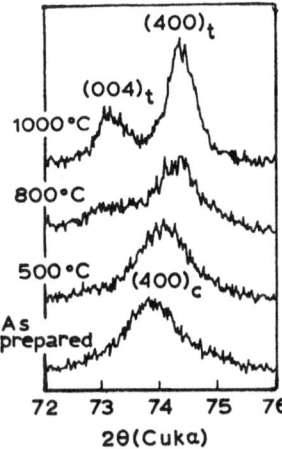

FIG. 19. X-ray diffraction patterns of PSZ (3 mol% Y_2O_3) powders calcined at various temperatures (from Ref. 20).

and refiltered. After four dispersion and filtration processes, the precipitate was dried at 120°C for 48 h.

The precipitates calcined at 240, 420, 620 and 820°C in air for 60 min at a heating rate of 10°C/min were also used as starting materials. Distilled water and solutions of 8 wt% KF, 30 wt% NaOH, 7, 15 and 30 wt% LiCl, and 10 wt% KBr were used as mineralizers. A Pt or Au tube 2·7 mm I.D., 0·15 mm thick and 32 mm long was filled with the starting material and solution, and treated under 100 MPa at 200–600°C for 24 h using a test-tube type pressure vessel. The temperature was measured by two Platinel (Au-Pd, Type K) thermocouples calibrated against the melting point of Zn (419.5°C) and controlled within ±3°C during a run. The pressure was measured by a calibrated Heise bourdon gauge and controlled within ±0·2 MPa. The vessel was then quenched into cold water. The products were washed in distilled water, dried and examined by X-ray powder diffractometry, transmission electron microscopy and energy-dispersive spectroscopy.

3.3.2. Results and discussion

The starting material dried at 120°C was shown to be amorphous (Fig. 20) by X-ray diffraction, and TG–DTA curves gave H_2O/ZrO_2 ratio of ≈ 2 for this material. This amorphous material was crystallized as monoclinic and/or tetragonal ZrO_2 by hydrothermal treatments.

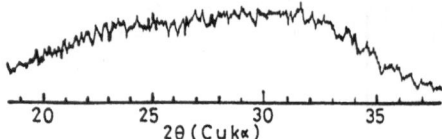

FIG. 20. X-ray diffraction pattern of amorphous hydrous zirconia used as starting material for hydrothermal crystallization, $ZrO_2 \cdot nH_2O$ ($n \sim 2$) (from Ref. 21).

Under hydrothermal conditions of 100 MPa, 300°C for 24 h, monoclinic ZrO_2 only formed in KF or NaOH solutions as shown in Fig. 21. In the case of KF solutions, only monoclinic ZrO_2 was obtained at temperatures of $\sim$200–600°C. The monoclinic ZrO_2 particles formed are homogeneous, non-agglomerated microcrystals with diameters of $\sim$20 nm (Fig. 22) from TEM observations.

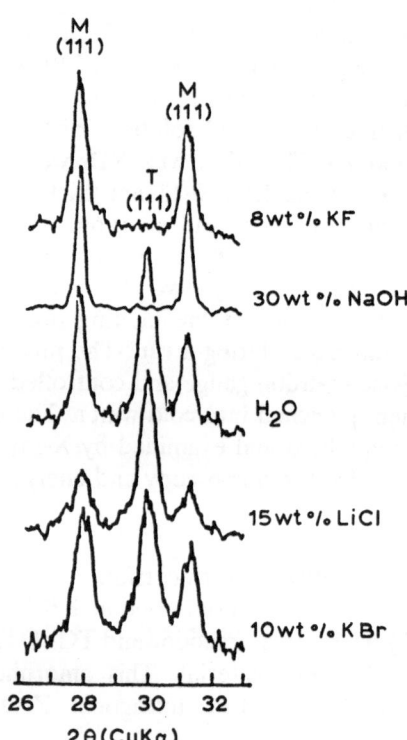

FIG. 21. X-ray diffraction patterns of the products crystallized hydrothermally from hydrous zirconia using various mineralizer solutions at 300°C under 100 MPa for 24 h (from Ref. 21).

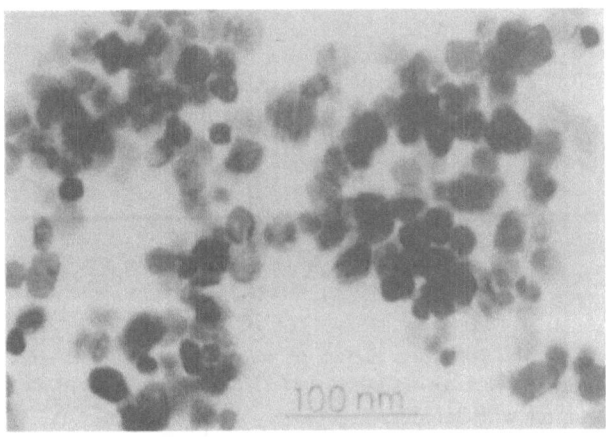

FIG. 22. Transmission electron micrograph of monoclinic ZrO_2 produced by hydrothermal crystallization at 400°C under 100 MPa for 24 h using 8 wt% KF solution (from Ref. 22).

The observed particle size is in good agreement with the crystallite size calculated from the half-width of the X-ray diffraction peak: 20 nm at 300°C, 21 nm at 400°C, as indicated in Fig. 23. This shows that the hydrothermally crystallized particles are single-grain microcrystals isolated from each other, in contrast to the agglomerated particles resulting from calcination of amorphous hydrous zirconia in air. Figure 24 shows that the crystallite size of monoclinic ZrO_2 increased slightly with increase of treatment temperature from 200 to 500°C, but increased rapidly above 500°C: i.e. to 70 nm at 550°C, and to $\gg 100$ nm at 600°C. This phenomenon appears to be due to the fact that the solubility of zirconia increased rapidly in KF solution and is consistent with the fact that the growth rate of monoclinic ZrO_2 crystal became significant above 600°C.

In the NaOH solution, the crystallites of monoclinic ZrO_2 were grown to 40 nm even at 300°C because of larger solubility than in KF solutions, as indicated in Table 3.

On the other hand, tetragonal ZrO_2 was crystallized in addition to a small amount of monoclinic ZrO_2 in the mineralizer solutions of H_2O, LiCl and KBr, as indicated in Table 3 and Fig. 18. The ratio of tetragonal ZrO_2/monoclinic ZrO_2 varied with the species and concentrations of the solutions and with treatment temperatures, but single-phase tetragonal ZrO_2 could not be obtained under the hydrothermal conditions studied. Figure 21 shows the tetragonal

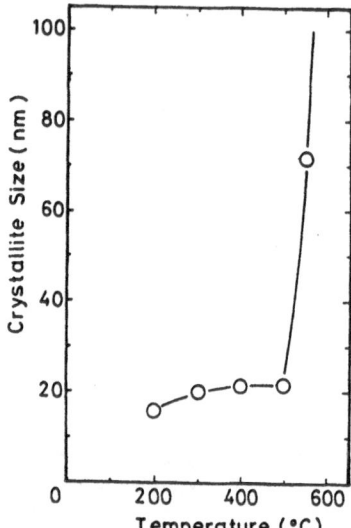

FIG. 23. Variation with temperature of crystallite size of monoclinic ZrO_2 produced by hydrothermal crystallization under 100 MPa for 24 h using 8 wt% KF solution (from Ref. 23).

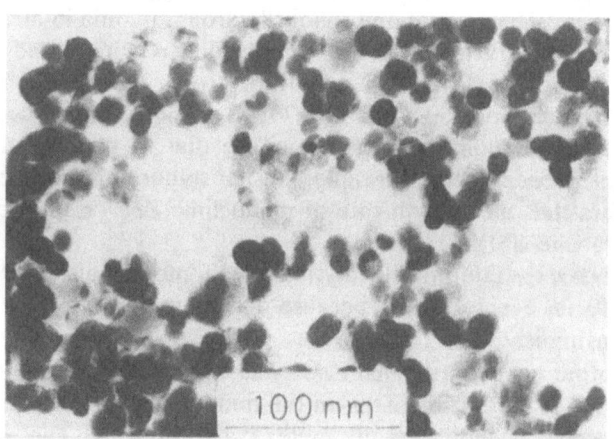

FIG. 24. Transmission electron micrograph of tetragonal (70%) and monoclinic (30%) ZrO_2 powders produced by hydrothermal crystallization at 300°C under 100 MPa for 24 h using 15 wt% LiCl solution (from Ref. 23).

TABLE 3
Phases present and crystallite size of products hydrothermally crystallized at 100 MPa for 24 h[a]

Mineralizer	Temperature (°C)	Average crystallite size (nm)	
		Tetragonal ZrO_2	Monoclinic ZrO_2
KF (8 wt%)	200	Not detected	16
KF (8 wt%)	300	Not detected	20
NaOH (30 wt%)	300	Not detected	40
H_2O	300	15	17
LiCl (15 wt%)	300	15	19
KBr (10 wt%)	300	13	15

[a] After Ref. 21.

(70%) and monoclinic (30%) ZrO_2 particles formed at 300°C in 15 wt% LiCl solution under 100 MPa for 24 h. The average crystallite size was calculated to be 15 nm for tetragonal ZrO_2 and 19 nm for monoclinic ZrO_2 in this product, as seen in Table 3. The particles in Fig. 24 seem to be single-crystal particles of tetragonal or monoclinic ZrO_2 because of the agreement between the particle size and the crystallite size.

3.4. Hydrothermal Synthesis
Synthesis of $LaCrO_3$ is described in this section [23].

3.4.1. Experimental
Chromium hydroxide was used as a starting material to improve the reactivity with La_2O_3. The chromium hydroxide was precipitated from 3 N chromic nitrate solution prepared from $Cr(NO_3)_2 \cdot 9H_2O$ by addition to 8 N NH_4OH solution. The precipitate was filtered, washed thoroughly with distilled water are then dried at 90°C for 24 h in air. The content of Cr_2O_3 in this hydroxide was analyzed by TG–DTA and ignition loss measurements. This chromium hydroxide was mixed with lanthanum oxide, of which the La_2O_3 content was also analyzed by TG–DTA and ignition loss measurements, to the exact composition of $La_2O_3 : Cr_2O_3 = 1:1$ (mol) by mixing thoroughly in an agate mortar with acetone.

About 0·1 g of the mixture was sealed with an electric arc in a Pt capsule of 2·7 mm I.D., 0·15 mm thick and 35 mm long. The capsule was heated in a test-tube type pressure vessel of Stellite 25 under a pressure of 100 MPa H_2O at 300–700°C for 3–24 h. The temperature was measured by calibrated Platinel thermocouples attached on the outside wall of the vessel.

In order to study the influence of atmosphere, additional experiments were carried out in which ~0·06 g chromium powder was added on top of the mixture in the capsule. Chromium reacts with H_2O to produce Cr_2O_3 and H_2 at high temperatures, so that a reducing atmosphere was created in the capsule.

3.4.2. Results

The chromium hydroxide prepared was fine-grained aggregate and was amorphous on the basis of the X-ray diffraction. The composition of this material appeared to be $Cr_2O_3 \cdot 6 \cdot 42H_2O$ from the ignition loss measurements. Based upon the X-ray diffraction, a small amount of $La(OH)_3$ was detected in the La_2O_3 used. The water contents of this material was determined to be $La_2O_3 \cdot 0 \cdot 95H_2O$ by ignition loss measurements and TG–DTA.

The product heated at 300°C was identified to be crystallized $La(OH)_3$ and a small amount of $LaCO_3OH$ by X-ray diffraction. Chromium hydroxide might exist still in the amorphous state, since no crystallized chromium compounds could be detected in this product. The formation of $LaCO_3OH$ indicates the contamination of starting La_2O_3 by CO_2 gas absorbed from atmosphere. Similar phenomena have often been observed in studies in the system $La_2O_3 – H_2O$ under hydrothermal conditions. The crystalline $LaCO_3OH$ is significantly stable under hydrothermal conditions, particularly at high pressures of 100 MPa or above in the presence of CO_2 components; even these were present in small amount in the system. $LaCO_3OH$ remained at 500°C and was present in trace amounts at 600°C, but has not been detected above 700°C. The product at 400°C was almost well-crystallized $LaCrO_3$, indicating reaction between $La(OH)_3$ and chromium hydroxide with the release of H_2O Scanning electron photomicrography of this product revealed fine-grained $LaCrO_3$ of about $0 \cdot 1 \ \mu m$ in diameter and hydrothermally grown $LaCO_3OH$ crystals of 1–$3 \ \mu m$ on the edge. As the reaction temperature was elevated, the amount and the size of $LaCO_3OH$ crystals decreased while the production of $LaCrO_3$ increased. Only well-crystallized $LaCrO_3$ could be observed in the product at 700°C. The $LaCrO_3$ crystals have homogeneous grain size of $0 \cdot 7 \pm 0 \cdot 2 \ \mu m$. The hydrothermally prepared

LaCrO$_3$ has orthorhombic symmetry with the lattice parameters $a = 5.477$ Å, $b = 5.511$ Å and $c = 7.758$ Å.

The results iindicate that formation of LaCrO$_3$ was completed in a short period of 3 h at temperatures as low as 700°C under hydrothermal conditions owing to the action of supercritical water as a mineralizer.

In the experiments with Cr powder, LaCrO$_3$ was also synthesized in the capsule at temperatures above 400°C. The products in the experiments with Cr powder were identical with those in the experiments without Cr powder. A reducing atmosphere created by the hydrogen produced by the reaction

$$2\,Cr + 3H_2O \rightarrow Cr_2O_3 + 3H_2$$

at temperatures above 600°C had little effect on the formation of LaCrO$_3$. These results indicate that LaCrO$_3$ is stable under reducing conditions, although oxygen-deficient Ca-doped LaCrO$_3$ has been prepared under hydrothermal conditions similar to those employed in the present study. The water was decomposed into hydrogen, which could permeate through the wall of the Pt capsule at high temperatures, so that a weight loss could be observed after hydrothermal treatment. The elimination of water contents, 12% or 18% at 600°C, corresponding to the observed weight losses, also showed no significant effects on the formation and crystal growth of LaCrO$_3$.

3.5. Hydrothermal Decomposition
Decomposition of ilmenite under hydrothermal conditions is given as an example [22].

3.5.1. Experimental
Natural ilmenite obtained from the beach sand deposit at Pulmoddai, Sri Lanka, was used in this investigation. The chemical analysis of the ilmenite showed the following:

TiO$_2$	53·61%
FeO	20·87%
Fe$_2$O$_3$	20·95%
MnO	0·98%

Ilmenite was powdered to pass through a 200-mesh sieve and was used in the hydrothermal experiments. 10 M KOH, 10 M NaOH and 10 M LiOH solutions were used. Known weights of ilmenite and alkali hydroxide solutions were introduced into the one-end sealed 2·7 mm I.D. and 35 mm long gold capsules and sealed hermetically. Both end-sealed capsules were placed in test-tube type hydrothermal equipment and the pressure was raised to the desired value. Then the temperature was raised while maintaining the pressure constant. At the end of the treatment the capsules were quenched in water.

3.5.2. Results

Ilmenite was completely decomposed under hydrothermal conditions in 10 M KOH solution at 500°C, when the pressure was between 25 and 35 MPa for 63 h of treatment. Ilmenite was stable above a pressure of 35 MPa at 500°C in 10 M KOH solution. The products formed by complete decomposition of ilmenite were magnetite ($Fe_{3-x}O_4$) and $K_2O·4TiO_2$. The lattice parameter of the magnetite ($Fe_{3-x}O_4$) formed was $a = 0·8467$ nm, which is larger than for stoichiometrically pure magnetite, for which $a = 0·8396(7)$ nm. The increase in lattice parameter is due to the substitution of Ti^{4+} in the magnetite lattice and formation of $Fe_{3-x}O_4$–Fe_2TiO_4 solid solution. On increasing the treatment temperature to 800°C while maintaining the pressure at 30 MPa, ilmenite decomposed completely in 10 M KOH solution with 24 h of treatment. The lattice parameters of the magnetite formed at 600, 700 and 800°C and at 30 MPa were determined and found to decrease with increasing temperature. The lattice parameter of magnetite formed at 800°C, 30 MPa and 24 h by the decomposition of ilmenite in 10 M KOH solution was $a = 0·8413(8)$ nm.

The decrease in lattice parameter of magnetite with increase of temperature under hydrothermal conditions was due to the exsolution of Ti^{4+} from $Fe_{3-x}O_4$–Fe_2TiO_4 solid solution to form nearly stoichiometrically pure magnetite.

The Curie points of the magnetite formed by the decomposition of ilmenite at 600, 700 and 800°C at 30 MPa and for 24 h using 10 M KOH solution were determined and found to increase with increase of temperature. The Curie point of the magnetite formed at 800°C was 573°C, whereas for stoichiometric magnetite the Curie point is 580°C.

The SEM micrograph of magnetite formed by the complete decomposition of ilmenite in 10 M KOH solution at 800°C, 30 MPa and 24 h is given in Fig. 25. The magnetite formed was of octahedral habit.

Ilmenite also completely decomposed in 10 M NaOH solution at 800°C, 30 MPa and 24 h of treatment. The products formed were magnetite ($Fe_{3-x}O_4$) and a non-stoichiometric sodium iron titanium

FIG. 25. Scanning electron micrograph of magnetite formed by the complete decomposition of ilmenite in 10 M KOH solution at 800°C, 30 MPa and 24 h of treatment (from Ref. 22).

oxide. The lattice parameter of the magnetite formed is given in Table 4 and it is larger than that for the magnetite formed in 10 M KOH and 10 M LiOH solutions under the same conditions.

The non-stoichiometric sodium iron titanium oxide formed was of orthorhombic structure with $a = 0.9083(6)$ nm, $b = 0.2954(1)$ nm and $c = 1.0718(3)$ nm.

This investigation clearly showed that nearly stoichiometrically pure magnetite can be synthesized by the decomposition of ilmenite in 10 M KOH solution at 800°C, 30 MPa and with 24 h of treatment.

TABLE 4

Lattice parameter of magnetite formed by the decomposition of ilmenite in different solutions at 800°C, 30 MPa and 24 h of treatment

Solution used	Lattice parameter of magnetite (nm)
10 M KOH	0.841 3(8)
10 M NaOH	0.844 7(5)
10 M LiOH	0.841 5(3)
Stoichiometric magnetite	0.839 6(7)

3.6. Hydrothermal Dehydration

Water is withdrawn from the hydrothermal solution through a spatially separated metal having a higher oxidation potential than that of hydrogen. Magnesium and zinc, which bring about the dehydration through formation of the corresponding hydrogen or oxide and hydrogen, have been used for this purpose (Fig. 26). In such cases the degree of fill should be as low as possible. Synthesis of $Mg_6Si_4O_{10}(OH)_8$ was performed using this method by Oota *et al.* [12].

3.6.1. Experimental

The starting materials were as follows: amorphous titanium dioxide; a mixture of ortho- and metatitanic acid obtained by the hydrolysis of titanium tetrachloride, and its composition dried at 110°C was $TiO_2 \cdot nH_2O$ ($n = 1 \cdot 1 – 1 \cdot 8$); Ti_2O_3 was prepared by calcining the mixture of the appropriate ratio of Ti to TiO_2 for 8 h at 1300°C in Ar gas; rutile-type TiO_2 was prepared by calcining TiO_2 of reagent grade; the potassium used was a commercial special grade of potassium hydroxide. Zinc powder or magnesium ribbon was used as the dehydration reagent.

The reaction vessels were stainless steel autoclaves of about 140-ml capacity and 3 cm inside diameter. A platinum tube containing nutrients and a solvent was placed in the autoclave. This tube was 10 cm long, 2·8 cm in diameter and open at the top for the dehydration. In the case of the dehydration method, another small tube

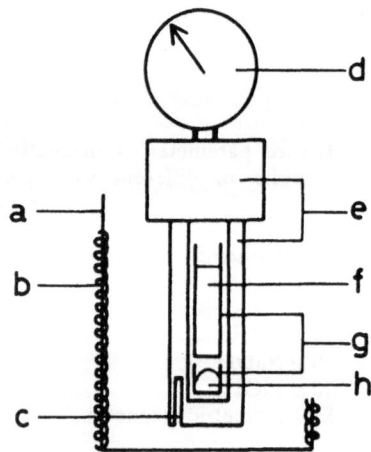

Fig. 26. Hydrothermal dehydration method: (a) furnace; (b) heater; (c) thermocouple well; (d) pressure gauge; (e) autoclave; (f) nutrient; (g) platinum tube; (h) dehydration agent (from Ref. 7).

containing the dehydration reagent was placed in the bottom of the autoclave. This assembly is shown in Fig. 26.

An adequate amount of $TiO_2 \cdot nH_2O$ (0.1–5.0 g) was admitted into the Pt-tube, and an appropriate amount of aqueous solution of KOH (25–60 ml) was added. The vessel was elevated to the desired temperature within 2 h and kept at this temperature for a given reaction period (1–50 h) in a furnace with a constant temperature controlling device; no temperature gradient was applied. After the vessel was cooled to room temperature, the solid products were washed out with water into a glass beaker, in which filmy long fibers tended to float or drift. The floating long fibers were removed by a mechanical process such as decantation, leaving a sponge-like residue in the beaker. Each part was then filtered.

3.6.2. Results
The dehydration method has been found to be effective at a lower percentage of fill for the formation of fibrous materials. In an autoclave filled with pure water to less than 32%, which is the critical fill, the liquid–gas meniscus moved down when the autoclave was heated to temperatures below 375°C, which is the critical temperature. Above the critical fill, therefore, a large degree of dehydration should be required for the present purpose. On the other hand, a liquid phase should exist even at 390°C in the present experiments, because the critical temperature of the solution is higher than that of pure water. Accordingly, the following runs were carried out at 20% fill.

In order to remove a part of the water under hydrothermal conditions, two kinds of metal—magnesium and zinc, which have higher ionization tendencies than hydrogen—were used as dehydration reagents. For 1 mole of water, about 1/9 mole of either metal was placed in the bottom of the autoclave. During the heating period, Mg and Zn metals were oxidized to $Mg(OH)_2$ and ZnO, respectively, and 1/10 of the amount of water was converted to the corresponding amount of hydrogen and oxide. As the reaction proceeded, the total pressure was increased slightly, because hydrogen gas more closely approaches ideal behavior than water. In the present experiments, the pressure of the hydrogen evolved was about 30 atm at room temperature after the reaction had finished.

Somewhat thicker fibers were obtained with the use of Mg and somewhat longer fibers with the use of Zn. This seems to be the result of differences in the ionization tendencies between Mg and Zn; as Mg has a higher ionization tendency than Zn, the dehydration with Mg proceeded so rapidly that alkaline concentration varied rapidly. The dehydration should proceed relatively slowly to produce long fibers, and Zn is considered as the favoured dehydration reagent.

3.7. Hydrothermal Anodic Oxidation

This method is one of hydrothermal oxidation with electrochemical reactions. $BaTiO_3$ was prepared by this method [14].

3.7.1. Experimental

The electrochemical cell and circuit arrangements for hydrothermal anodic oxidation are shown schematically in Fig. 27. Anodic oxidation was performed in various electrolytes: $Ba(NO_3)_2$ $(0\cdot01-0\cdot5\,\text{N})$, $0\cdot1\,\text{N}\,BaCl_2$, and pure H_2O. The quantity of electrolyte was 200 ml and the area of working electrode exposed to electrolyte was $2\cdot1\,\text{cm}^2$ for treatment. As a counterelectrode (cathode), a platinum plate (9 mm × 30 mm × 0·15 mm) was used. The distance between the working electrode and the platinum counterelectrode was 3·0 cm.

Anodic oxidation was carried out galvanostatically with an applied current density between 10 and 100 mA cm^{-2} using a galvanostat in an electrolytic autoclave under the saturation vapour pressure of the

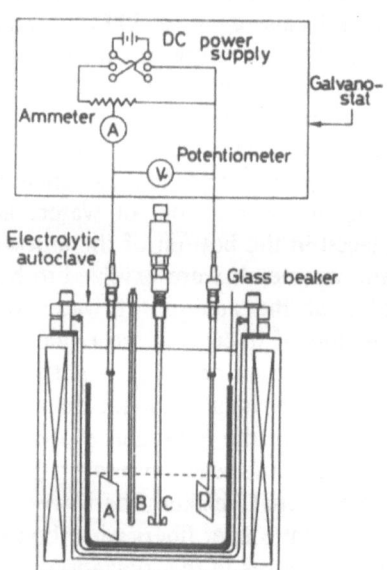

FIG. 27. Schematic illustration of the electrochemical cell and circuit arrangements for anodic oxidation under hydrothermal conditions. A, counterelectrode (platinum plate), cathode; B, thermocouple; C, stirrer; D, working electrode (titanium plate), anode (from Ref. 14).

electrolyte between room temperature and 250°C. A silica glass beaker was set to prevent the autoclave wall from corrosion by contact with the electrolyte solution.

Titanium plates (99·9% pure) 0·1 mm thick were used as working electrodes (anodes). Prior to each experiment the sample was degreased in acetone, chemically etched with 6 N HCl, and then cleaned.

3.7.2. Results

Anatase was the only product of anodic oxidation of titanium plate in H_2O at 200–250°C. Similarly, when a 0·1 N $BaCl_2$ solution was used as electrolyte solution, only anatase powders were produced, even at 250°C.

Anodic oxidation of titanium metal plate in 0·1 N $Ba(NO_3)_2$ solution also yielded only TiO_2 (anatase) powders with low crystallinity below 200°C, regardless of the applied current density. However, when the treatment temperature was raised to 250°C, perovskite $BaTiO_3$ (cubic or pseudocubic phase) began to form.

On the other hand, $BaTiO_3$ was formed even at 200°C in solutions with higher concentrations of barium (0·5 N $Ba(NO_3)_2$), although a small amount of $BaCO_3$ or TiO_2 (anatase) formed in addition to $BaTiO_3$ at this temperature. Figure 28 shows the X-ray diffraction patterns of the powders produced and of the surfaces of working

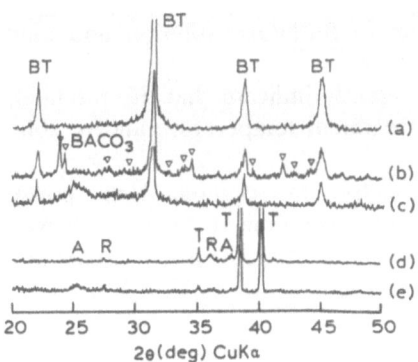

Fig. 28. X-ray diffraction patterns of the powders produced and surfaces of working electrodes in 0·5 N $Ba(NO_3)_2$ solution under various conditions. (a) Powder, 250°C, 100 mA cm^{-2}. (b) Powder, 200°C, 100 mA cm^{-2}. (c) Powder, 200°C, 50 mA cm^{-2}. (d) Working electrode, 250°C, 100 mA cm^{-2}. (e) Working electrode, 200°C, 100 mA cm^{-2}. BT = $BaTiO_3$; A = anatase; R = rutile; T = titanium (from Ref. 14).

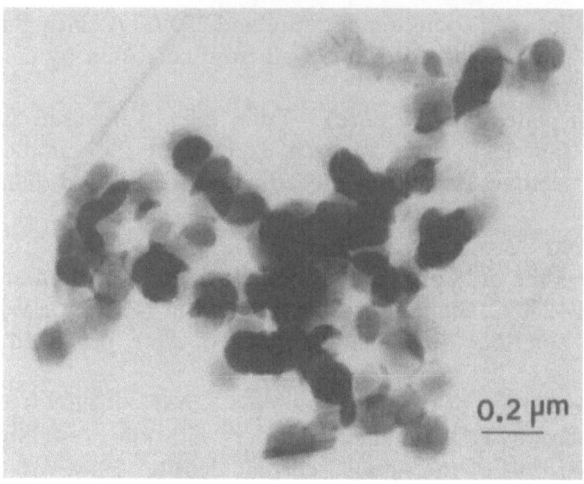

Fig. 29. Transmission electron micrograph of the BaTiO₃ powders obtained after hydrothermal–electrochemical treatment of titanium plate in 0·5 N Ba(NO₃)₂ solution at 250°C (from Ref. 14).

electrodes after hydrothermal anodic oxidation treatments of titanium metal plates in $0·5 N Ba(NO_3)_2$ solution. At 250°C, pure $BaTiO_3$ powders could be obtained, whereas the surface of the working electrode was always only TiO_2 after hydrothermal anodic oxidation of titanium metal plate in $Ba(NO_3)_2$ solution under all the conditions studied.

These results apparently indicate that the production of $BaTiO_3$ is dependent on the treatment temperature and/or solution species and concentration.

A transmission electron micrograph of the powder product produced in $0·5 N Ba(NO_3)_2$ solution at 250°C is shown in Fig. 29. The products consisted of submicrometer fine powders of cubic or pseudocubic $BaTiO_3$.

3.8. RESA
Reactive-electrode submerged arc (RESA) is a new process for making fine (10–1000 nm) oxide and non-oxide powders. This process was reported by Kumar and Roy [15,16].

3.8.1. Experimental
The technique utilizes two metal or conducting electrodes, which are immersed in a suitable dielectric fluid that reacts with metal. A spark

provides an extremely high temperature and high pressure for a short duration, resulting in vaporization of the electrode and the surrounding dielectric fluid and reaction within the 'bubble' which forms. A schematic of the electrode region during the passage of current in the arc is depicted in Figs 30 and 31.

The vapors of the metal and dielectric materials remain confined within the bubble. Reaction occurs between the various vapor species or between solids and vapor, and the reaction products in the form of a fine smoke are quenched in the surrounding liquid-phase dielectric, yielding a colloidal sol of spherically shaped particles.

The principle of the apparatus is rather simple and the application of normal line voltages (~230 V) across thin metal foils immersed in water or mineral oil is sufficient to cause a spark capable of producing colloidal particles. The sophistication of such an apparatus requires control of the voltage and current parameters, the spark gap, fluid medium, and so forth. The first efforts in apparatus construction were directed towards obtaining a reasonably constant spark between similar metals. Subsequently, the apparatus was refined and interfaced with a microcomputer to monitor and adjust the spark and electrode feed rate.

An arc is struck between two metal electrodes typically 3–6 mm in

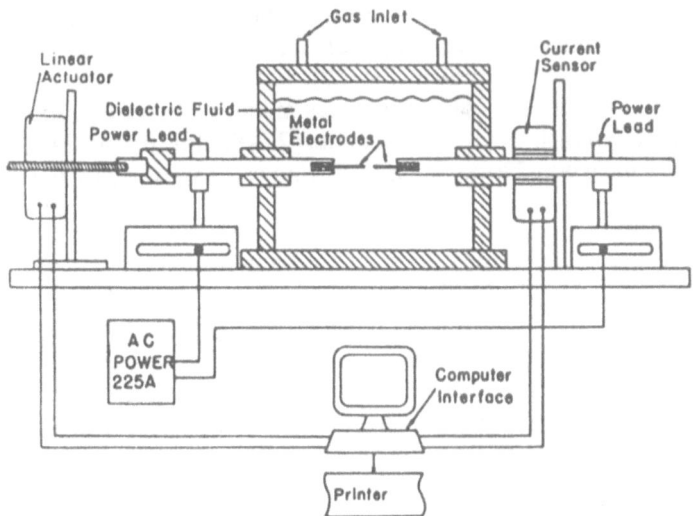

FIG. 30. Schematic of microprocessor-controlled RESA apparatus for fine powder preparation (from Ref. 15).

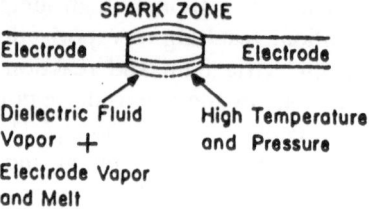

SPARK ZONE

FIG. 31. Schematic of the spark zone at the instant of the spark between metal electrodes submerged in a dielectric liquid (from Ref. 15).

Electrode | High Temperature and Pressure

Dielectric Fluid Vapor + Electrode Vapor and Melt

diameter that are immersed in approximately 1000 ml of a liquid, typically deionized water. To keep the arc running as material is consumed, a linear actuator capable of 0·025-mm steps in either direction controls the movement of one electrode with respect to the other (fixed) electrode.

3.8.2. Results

Zirconium rod (2 mm in diameter) under conditions of 70 V and 40–60 A was used to produce zirconium oxide. This process produced 10–100-nm powder. Average particle size was 50 nm. The process was able to make Al_2O_3, ZrO_2, Ti oxides, Cr_2O_3, ZnO, and other oxides.

3.9. Hydrothermal Mechanochemical Reactions [17]

3.9.1. Experimental

Apparatus for this method used in the author's laboratory consists of an autoclave with a Teflon propeller. This is an autoclave with an attrition ball-mixing equipment (Fig. 32). Ba^2/Fe^3 (1:8 of $Ba(OH)_2$ and $FeCl_3$) solution with NaOH solution was made. OH concentration was constant at 3 mol/litre at 200°C under 2 MPa for up to 24 h. The number of balls was 200 to 700 and stirring rates were up to 107 rpm.

3.9.2. Results

Figure 33 shows the results of X-ray diffraction of the samples: (a) is the starting material, (b) is hydrothermally synthesized at 200°C, under 2 MPa for 4 h without mixing, and (c) is synthesized by hydrothermal attrition mixing at 200°C, at 37 rpm using 200 balls. Although the starting material is amorphous, barium-ferrite was formed in both hydrothermal synthesis and hydrothermal attrition mixing. After a detailed study of the results of X-ray diffraction, however, a weak halo was observed at the low-angle side. This suggests that amorphous elements in the starting material remained in the sample without being crystallized.

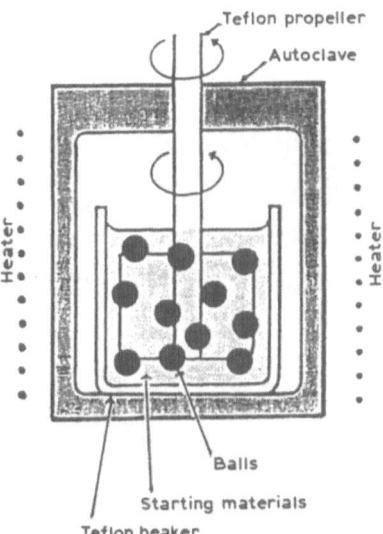

FIG. 32. Experimental apparatus for hydrothermal mechanochemical reactions. (from Ref. 17).

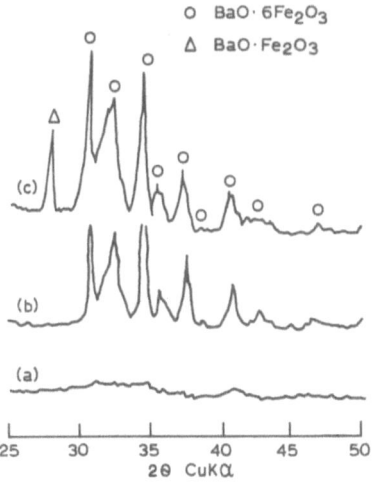

FIG. 33. X-ray diffraction profiles of (a) starting material; (b) material fabricated at 200°C under 2 MPa for 4 h, without rotation; and (c) material fabricated at 200°C for 4 h, using 200 balls, 37 rpm (from Ref. 17).

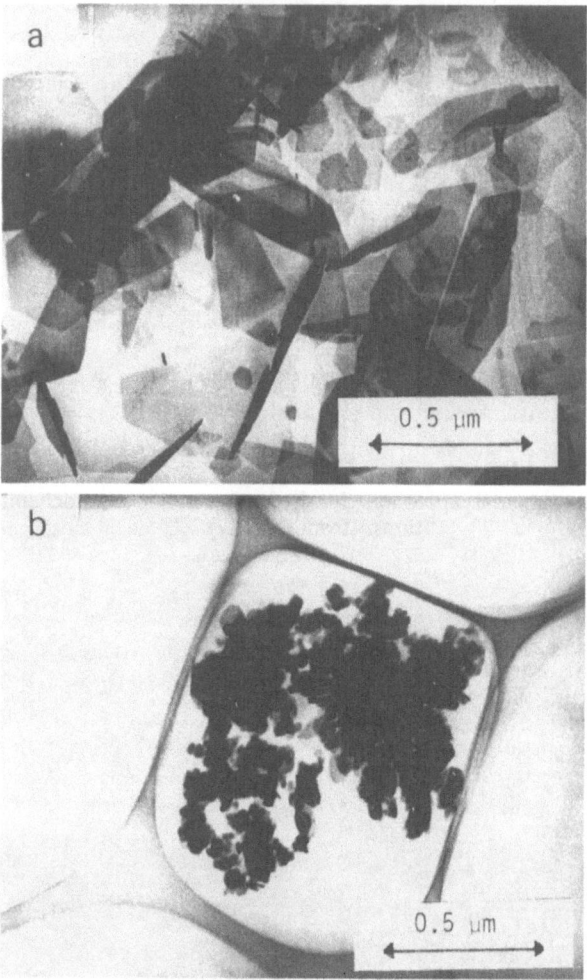

FIG. 34. Transmission electron micrographs of (a) material fabricated at 200°C under 2 MPa for 4 h, without rotation, and (b) material fabricated at 200°C under 2 MPa for 4 h, using 700 balls, 107 rpm (from Ref. 17).

Nevertheless, as is seen in Fig. 33, BaO·Fe$_2$O$_3$ was formed as a secondary phase. When BaO·Fe$_2$O$_3$ was quantitatively analyzed by X-ray diffraction, the rate of formation was found to increase with increased number of balls and faster rotation rate, reaching a constant level at about 3 wt%.

Figures 34(a) and 34(b) show TEM photos of barium–ferrite formed without balls, and formed with 700 balls and rotation speed of 107 rpm, held for 4 h, respectively. Barium–ferrite formed is hexagonal plate-like, its mean particle diameter is 170 nm, and its thickness is 10 nm (aspect ratio = 17). Here, needle-like crystals seen in Fig. 34(a) are in fact plate-like crystals seen on edge. From Fig. 34(b), the mean particle size is 40 nm, and the thickness is 10 nm (aspect ratio = 4); this suggests that the method using balls produces smaller particle size.

Attrition mixing under hydrothermal conditions affects mainly the rate of nucleation; when the number of balls or the rotation speed is increased, the nucleation rate is expected to increase.

4. SUMMARY

This chapter has described the preparation of fine powders by hydrothermal processing. There are several ways of preparing fine powders under hydrothermal conditions. This process is one of the best approaches for making ideal powders for ceramics. However, morphology, particle size, size of crystallite, degree of agglomeration, distribution of particle size, etc., are different from process to process. Powders are one of key materials for making advanced ceramics.

REFERENCES

1. G. W. Morey, Hydrothermal synthesis, *J. Am. Ceram. Soc.*, **36**(9), 279–285 (1953).
2. R. Roy and O. F. Tuttle, Investigations under hydrothermal conditions. In *Physics and Chemistry of the Earth*, ed. L. H. Ahrens, K. Pankama and S. K. Runcorn, Pergamon Press, Oxford, Vol. 1, 1955, pp. 138–80.
3. R. A. Laudise and J. W. Nielsen, Hydrothermal crystal growth, *Solid-State Phys.*, **12**, 149–222 (1961).
4. R. A. Laudise, The growth of single crystals. In *Hydrothermal Growth*, Prentice Hall, Englewood Cliffs, N.J., 1970, pp. 275–93.
5. R. A. Laudise, Hydrothermal growth. In *Crystal Growth: An Introduction*, ed. P. Hartman. North-Holland, Amsterdam, 1973, pp. 163–97.

6. A. Rabenau, The role of hydrothermal synthesis in preparative chemistry, *Angew. Chem. Int. Ed. Engl.*, **24**, 1026–1040 (1985).
7. A. Rabenau, The role of hydrothermal synthesis in materials science, *J. Mater. Educ.*, **10**(5), 543–592 (1988).
8. A. A. Ballman and R. A. Laudise, Hydrothermal growth. In *The Art and Science of Growing Crystals*, ed. J. J. Gilman, Wiley, New York, 1963, pp. 231–51.
9. L. N. Demianets, V. A. Kuznetzov and A. N. Lobachev, Growth and synthesis in hydrothermal solutions. In *Modern Crystallography III* ed. A. A. Cherno, Springer-Verlag, Berlin, 1984, pp. 380–406.
10. S. Sōmiya, Hydrothermal synthesis of electronic and magnetic materials, *Zairyo Kagaku*, **13**, 55–62, 143–51 (1976).
11. J. Asahara, K. Nagai and S. Harada, Synthetic quartz crystals by large autoclaves—Their quality and characterization. In *Proceedings of the First International Symposium on Hydrothermal Reactions*, ed. S. Sōmiya, Gakujutsu Bunken Fukyu Kai, Tokyo, 1983, pp. 430–41.
12. T. Oota, H. Saito and I. Yamai, Synthesis of potassium hexatitanate fibers by the hydrothermal dehydration, *J. Crystal Growth*, **46**, 331–8 (1979).
13. S. E. Yoo, M. Yoshimura and S. Sōmiya, Preparation of $BaTiO_3$ and $LiNbO_3$ powders by hydrothermal anodic oxidation. In *Sintering '87*, ed. S. Sōmiya, M. Shimada, M. Yoshimura and R. Watanabe. Elsevier Applied Science Publishers. London, UK, 1989, pp. 108–13.
14. S. E. Yoo, M. Yoshimura and S. Sōmiya, Direct preparation of $BaTiO_3$ powders from titanium metal by anodic oxidation under hydrothermal conditions. *J. Mat. Sci. Letters*, **8**, 530–2 (1989).
15. A. Kumar and R. Roy, RESA—A wholly new process for fine oxide powder preparation, *J. Mater. Res.*, **3**(6), 1373–7 (1988).
16. A. Kumar and R. Roy, Reactive-electrode submerged arc process for producing fine non-oxide powders, *J. Am. Ceram. Soc.*, **72**(2), 354–6 (1989).
17. M. Yoshimura, N. Kubotera, T. Norma and S. Sōmiya, Synthesis of Ba-Ferrite fine particle by hydrothermal attrition mixing. *J. Ceram. Soc. Int. Ed.*, **97**, 14–19 (1989).
18. M. Yoshimura, H. Ōhira and S. Sōmiya, Hydrothermal oxidation of titanium metal, *Yogyo Kyokai Shi*, **93**(17), 359–63 (1985).
19. S. Sōmiya, M. Yoshimura, Z. Nakai, K. Hishinuma and T. Kumaki, Microstructure development of hydrothermal powders and ceramics. In *Ceramic Microstructure '86*, ed. J. A. Pask and H. G. Evans, Plenum Press, 1987, pp. 465–74.
20. K. Hishinuma, T. Kumaki, Z. Nakai, M. Yoshimura and S. Sōmiya, Characterization of Y_2O_3–ZrO_2 powders synthesized under hydrothermal conditions. *Advances in Ceramics*, **24**, 201–9 (1988).
21. M. Yoshimura and S. Sōmiya, Fine zirconia powders by hydrothermal processing, *Rept. Res. Lab. Eng. Mat. Tokyo Institute of Technology*, No. 9, 53–64 (1984).
22. M. G. M. U. Ismail, M. Yoshimura and S. Sōmiya, Synthesis of magnetite using ilmenite under hydrothermal conditions. *J. Mat. Sci. Letters* **1**(1), 19–21 (1985).

23. M. Yoshimura, S. T. Song and S. Sōmiya, Hydrothermal synthesis and sintering of LaCrO₃. In *Ferrites,* ed. H. Watanabe, S. Iida and M. Sugimoto, Academic Publications, 1981, pp. 429–32.
24. S. Sōmiya and M. Yoshimura, Hydrothermal preparation of fine powders. In *Fundamental Structural Ceramics,* ed. S. Sōmiya and R. C. Bradt, Terra Sci. Pub. Co., 1987, pp. 11–29.

APPENDIX: SSG DEFINITIONS

1. *Single phase gel* refers to the single solid phase in the gel which consists then of one finely divided mechanically plastic solid phase with water (or other liquid) in the interstices.
2. A di-phasic gel is a material which consists of two chemically or structurally distinct phases (in the Gibbsonian sense), each in the colloidal size range, which form a plastic solid by including some liquid in interstices between the solid particles.
3. Xerogels are dried or dehydrated gels, i.e. with the water (or liquid) removed from a diphasic gel. Aerogels are a special subset of xerogels where by using critical point drying the *microstructure of the original* gel is preserved.
4. Di-phasic xerogel is a dried di-phasic gel.

These are definitions made by Professor Rustum Roy relating to the preparation of fine powders (Roy, R., 1988, pers. comm.).

Index